D0999954

# Visual Data Mining

# Visual Data Mining

## The VisMiner Approach

**RUSSELL K. ANDERSON**
VisTech, USA

A John Wiley & Sons, Ltd., Publication

This edition first published 2013
© 2013 John Wiley & Sons, Ltd.

*Registered office*
John Wiley & Sons, Ltd., The Atrium, Southern Gate, Chichester, West Sussex, PO19 8SQ, United Kingdom

For details of our global editorial offices, for customer services and for information about how to apply for permission to reuse the copyright material in this book please see our website at www.wiley.com.

The right of the author to be identified as the author of this work has been asserted in accordance with the Copyright, Designs and Patents Act 1988.

All rights reserved. No part of this publication may be reproduced, stored in a retrieval system, or transmitted, in any form or by any means, electronic, mechanical, photocopying, recording or otherwise, except as permitted by the UK Copyright, Designs and Patents Act 1988, without the prior permission of the publisher.

Wiley also publishes its books in a variety of electronic formats. Some content that appears in print may not be available in electronic books.

Designations used by companies to distinguish their products are often claimed as trademarks. All brand names and product names used in this book are trade names, service marks, trademarks or registered trademarks of their respective owners. The publisher is not associated with any product or vendor mentioned in this book. This publication is designed to provide accurate and authoritative information in regard to the subject matter covered. It is sold on the understanding that the publisher is not engaged in rendering professional services. If professional advice or other expert assistance is required, the services of a competent professional should be sought.

*Library of Congress Cataloging-in-Publication Data*

Anderson, Russell K.
Visual data mining : the VisMiner approach / Russell K. Anderson.
    p. cm.
  Includes index.
  ISBN 978-1-119-96754-5 (cloth)
  1.  Data mining.  2.  Information visualization.  3.  VisMiner (Electronic resource)   I. Title.
  QA76.9.D343A347 2012
  006.3'12–dc23                                                        2012018033

A catalogue record for this book is available from the British Library.

ISBN: 9781119967545

Set in 10.25/12pt Times by Thomson Digital, Noida, India.
Printed and bound in Singapore by Markono Print Media Pte Ltd.

# Contents

# Preface

VisMiner was designed to be used as a data mining teaching tool with application in the classroom. It visually supports the complete data mining process – from dataset preparation, preliminary exploration, and algorithm application to model evaluation and application. Students learn best when they are able to visualize the relationships between data attributes and the results of a data mining algorithm application.

This book was originally created to be used as a supplement to the regular textbook of a data mining course in the Marriott School of Management at Brigham Young University. Its primary objective was to assist students in learning VisMiner, allowing them to visually explore and model the primary text datasets and to provide additional practice datasets and case studies. In doing so, it supported a complete step-by-step process for data mining.

In later revisions, additions were made to the book introducing data mining algorithm overviews. These overviews included the basic approach of the algorithm, strengths and weaknesses, and guidelines for application. Consequently, this book can be used both as a standalone text in courses providing an application-level introduction to data mining, and as a supplement in courses where there is a greater focus on algorithm details. In either case, the text coupled with VisMiner will provide visualization, algorithm application, and model evaluation capabilities for increased data mining process comprehension.

As stated above, VisMiner was designed to be used as a teaching tool for the classroom. It will effectively use all display real estate available. Although the complete VisMiner system will operate within a single display, in the classroom setting we recommend a dual display/projector setting. From experience, we have also found that students using VisMiner also prefer the dual display setup. In chatting with students about their experience with VisMiner, we found that they would bring their laptop to class, working off a single display, then plug in a second display while solving problems at home.

An accompanying website where VisMiner, datasets, and additional problems may be downloaded is available at www.wiley.com/go/visminer.

# Acknowledgments

The author would like to thank the faculty and students of the Marriott School of Management at Brigham Young University. It was their testing of the VisMiner software and feedback for drafts of this book that has brought it to fruition. In particular, Dr. Jim Hansen and Dr. Douglas Dean have made extraordinary efforts to incorporate both the software and the drafts in their data mining courses over the past three years.

In developing and refining VisMiner, Daniel Link, now a PhD student at the University of Southern California, made significant contributions to the visualization components. Dr. Musa Jafar, West Texas A&M University provided valuable feedback and suggestions.

Finally, thanks go to Charmaine Anderson and Ryan Anderson who provided editorial support during the initial draft preparation.

# 1

# Introduction

Data mining has been defined as the search for useful and previously unknown patterns in large datasets. Yet when faced with the task of mining a large dataset, it is not always obvious where to start and how to proceed. The purpose of this book is to introduce a methodology for data mining and to guide you in the application of that methodology using software specifically designed to support the methodology. In this chapter, we provide an overview of the methodology. The chapters that follow add detail to that methodology and contain a sequence of exercises that guide you in its application. The exercises use VisMiner, a powerful visual data mining tool which was designed around the methodology.

## Data Mining Objectives

Normally in data mining a mathematical model is constructed for the purpose of **prediction** or **description**. A model can be thought of as a virtual box that accepts a set of inputs, then uses that input to generate output.

Prediction modeling algorithms use selected input attributes and a single selected output attribute from your dataset to build a model. The model, once built, is used to predict an output value based on input attribute values. The dataset used to build the model is assumed to contain historical data from past events in which the values of both the input and output attributes are known. The data mining methodology uses those values to construct a model that best fits the data. The process of model construction is sometimes referred to as **training**. The primary objective of model construction is to use the model for predictions in the future using known input attribute values when the value

*Visual Data Mining: The VisMiner Approach*, First Edition. Russell K. Anderson.
© 2013 John Wiley & Sons, Ltd. Published 2013 by John Wiley & Sons, Ltd.

of the output attribute is not yet known. Prediction models that have a categorical output are known as **classification** models. For example, an insurance company may want to build a classification model to predict if an insurance claim is likely to be fraudulent or legitimate.

Prediction models that have numeric output are called **regression** models. For example, a retailer may use a regression model to predict sales for a proposed new store based on the demographics of the store. The model would be built using data from previously opened stores.

One special type of regression modeling is **forecasting**. Forecasting models use time series data to predict future values. They look at trends and cycles in previous periods in making the predictions for future time periods.

Description models built by data mining algorithms include: **cluster**, **association**, and **sequence** analyses.

Cluster analysis forms groupings of similar observations. The clusterings generated are not normally an end process in data mining. They are frequently used to extract subsets from the dataset to which other data mining methodologies may be applied. Because the behavioral characteristics of sub-populations within a dataset may be so different, it is frequently the case that models built using the subsets are more accurate than those built using the entire dataset. For example, the attitude toward, and use of, mass transit by the urban population is quite different from that of the rural population.

Association analysis looks for sets of items that occur together. Association analysis is also known as market basket analysis due to its application in studies of what consumers buy together. For example, a grocery retailer may find that bread, milk, and eggs are frequently purchased together. Note, however, that this would not be considered a real data mining discovery, since data mining is more concerned with finding the unexpected patterns rather than the expected.

Sequence analysis is similar to association analysis, except that it looks for groupings over time. For example, a women's clothing retailer may find that within two weeks of purchasing a pair of shoes, the customer may return to purchase a handbag. In bioinformatics, DNA studies frequently make use of sequence analysis.

## Introduction to VisMiner

**VisMiner** is a software tool designed to visually support the entire data mining process. It is intended to be used in a course setting both for individual student use and classroom lectures when the processes of data mining are presented. During lectures, students using VisMiner installed on desktop, laptop, tablet computers, and smart phones are able to actively participate with the instructor as datasets are analyzed and the methodology is examined.

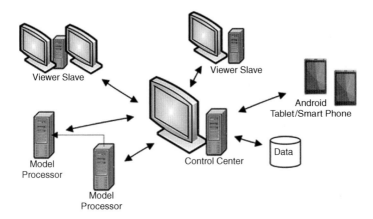

**Figure 1.1**   VisMiner Architecture

The architecture of VisMiner is represented in Figure 1.1. It consists of four main components:

- the **Control Center**, which manages the datasets, starts and stops the modelers and viewers, and coordinates synchronization between viewers

- **VisSlave** and **ModelSlave** which establish the connections between a slave computer and the Control Center

- the **modelers** that execute the sophisticated data mining algorithms

- the **viewers** that present interactive visualizations of the datasets and the models generated using the datasets.

As evidenced by Figure 1.1, VisMiner may run on one or more computers. The primary computer runs the Control Center. Computers that will present visualizations should run VisSlave; computers that will be used for back-end processing should run ModelSlave. In the full configuration of VisMiner, there should be just one instance of the Control Center executing, and as many instances of VisSlave and ModelSlave as there are computers available for their respective purposes. If there is only one computer, use it to run all three applications.

## The Data Mining Process

Successful data mining requires a potentially time-consuming and methodical process. That's why they call it "mining". Gold prospectors don't buy their gear, head out and discover gold on the first day. For them it takes months or even years of search. The same is true with data mining. It takes work, but hopefully not months or years.

In this book, we present a methodology. VisMiner is designed to support and streamline the methodology. The methodology consists of four steps:

- **Initial data exploration** – conduct an initial exploration of the data to gain an overall understanding of its size and characteristics, looking for clues that should be explored in more depth.

- **Dataset preparation** – prepare the data for analysis.

- **Algorithm application** – select and apply data mining algorithms to the dataset.

- **Results evaluation** – evaluate the results of the algorithm applications, assessing the "goodness of fit" of the data to the algorithm results and assessing the nature and strengths of inputs to the algorithm outputs.

These steps are not necessarily sequential in nature, but should be considered as an iterative process progressing towards the end result – a complete and thorough analysis. Some of the steps may even be completed in parallel. This is true for "Initial data exploration" and "dataset preparation". In VisMiner for example, interactive visualizations designed primarily for the initial data exploration also support some of the dataset preparation tasks.

In the sections that follow, we elaborate on the tasks to be completed in each of the steps. In later chapters, problems and exercises are presented that guide you through completion of these tasks using VisMiner. Throughout the book, reference is made back to the task descriptions introduced here. It is suggested that as you work through the problems and exercises, you refer back to this list. Use it as a reminder of what has and has not been completed.

## Initial data exploration

The primary objective of initial data exploration is to help the analyst gain an overall understanding of the dataset. This includes:

- **Dataset size and format** – Determine the number of observations in the dataset. How much space does it occupy? In what format is it stored? Possible formats include tab or comma delimited text files, fixed field text files, tables in a relational database, and pages in a spreadsheet. Since most datasets stored in a relational database are encoded in the proprietary format of the database management system used to store the data, check that you have access to software that can retrieve and manipulate the content. Look also at the number of tables containing data of interest. If found in multiple tables, determine how they are linked and how they might be joined.

- **Attribute enumeration** – Begin by browsing the list of attributes contained in the dataset and the corresponding types of each attribute. Understand what each attribute represents or measures and the units in which it is encoded. Look for identifier or key attributes – those that uniquely identify observations in the dataset.

- **Attribute distributions** – For numeric types, determine the range of values in the dataset, then look at the shape and symmetry or skew of the distribution. Does it appear to approximate a normal distribution or some other distribution? For nominal (categorical) data, look at the number of unique values (categories) and the proportion of observations belonging to each category. For example, suppose that you have an attribute called *CustomerType*. The first thing that you want to determine is the number of different *CustomerTypes* in the dataset and the proportions of each.

- **Identification of sub-populations** – Look for attribute distributions that are **multimodal** – that is distributions that have multiple peaks. When you see such distributions, it indicates that the observations in the dataset are drawn from multiple sub-populations with potentially different distributions. It is possible that these sub-populations could generate very different models when submitted in isolation to the data mining algorithms as compared to the model generated when submitting the entire dataset. For example, in some situations the purchasing behavior of risk-taking individuals may be quite different from those that are risk averse.

- **Pattern search** – Look for potentially interesting and significant relationships (or patterns) between attributes. If your data mining objective is the generation of a prediction model, focus on relationships between your selected output attribute and attributes that may be considered for input. Note the type of the relationship – linear or non-linear, direct or inverse. Ask the question, "Does this relationship seem reasonable?" Also look at relationships between potential input attributes. If they are highly correlated, then you probably want to eliminate all but one as you conduct in-depth analyses.

## Dataset preparation

The objective of dataset preparation is to change or morph the dataset into a form that allows the dataset to be submitted to a data mining algorithm for analysis. Tasks include:

- **Observation reduction** – Frequently there is no need to analyze the full dataset when a subset is sufficient. There are three reasons to reduce the observation count in a dataset.

  o The amount of time required to process the full dataset may be too computationally intensive. An organization's actual production database may have millions of observations (transactions). Mining of the entire dataset may be too time-consuming for processing using some of the available algorithms.

○ The dataset may contain sub-populations which are better mined independently. At times, patterns emerge in sub-populations that don't exist in the dataset as a whole.

○ The level of detail (**granularity**) of the data may be more than is necessary for the planned analysis. For example, a sales dataset may have information on each individual sale made by an enterprise. However, for mining purposes, sales information summarized at the customer level or other geographic level, such as zip code, may be all that is necessary.

Observation reduction can be accomplished in three ways:

○ extraction of sub-populations

○ sampling

○ observation aggregation.

- **Dimension reduction** – As dictated by the "**curse of dimensionality**", data becomes more **sparse** or spread out as the number of dimensions in a dataset increases. This leads to a need for larger and larger sample sizes to adequately fill the data space as the number of dimensions (attributes) increases. In general, when applying a dataset to a data mining algorithm, the fewer the dimensions the more likely the results are to be statistically valid. However, it is not advisable to eliminate attributes that may contribute to good model predictions or explanations. There is a trade-off that must be balanced.

To reduce the dimensionality of a dataset, you may selectively remove attributes or arithmetically combine attributes.

Attributes should be removed if they are not likely to be relevant to an intended analysis or if they are redundant. An example of an irrelevant attribute would be an observation identifier or key field. One would not expect a customer number, for example, to contribute anything to the understanding of a customer's purchase behavior. An example of a redundant attribute would be a measure that is recorded in multiple units. For example, a person's weight may be recorded in pounds and kilograms – both are not needed.

You may also arithmetically combine attributes with a formula. For example, in a "homes for sale" dataset containing *price* and *area* (square feet) attributes, you might derive a new attribute "price per square foot" by dividing *price* by *area*, then eliminating the *price* and *area* attributes.

A related methodology for combining attributes to reduce the number of dimensions is **principal component analysis**. It is a mathematical procedure in which a set of correlated attributes are transformed into a potentially smaller and uncorrelated set.

- **Outlier detection** – Outliers are individual observations whose values are very different from the other observations in the dataset. Normally, outliers are erroneous data resulting from problems during data capture, data entry, or data encoding and should be removed from the dataset as they will distort results. In some cases, they may be valid data. In these cases, after verifying the validity of the data, you may want to investigate further – looking for factors contributing to their uniqueness.

- **Dataset restructuring** – Many of the data mining algorithms require a single tabular input dataset. A common source of mining data is transactional data recorded in a relational database, with data of interest spread across multiple tables. Before processing using the mining algorithms, the data must be joined in a single table. In other instances, the data may come from multiple sources such as marketing research studies and government datasets. Again, before processing the data will need to be merged into a single set of tabular data.

- **Balancing of attribute values** – Frequently a classification problem attempts to identify factors leading to a targeted anomalous result. Yet, precisely because the result is anomalous, there will be few observations in the dataset containing that result if the observations are drawn from the general population. Consequently, the classification modelers used will fail to focus on factors indicating the anomalous result, because there just are not enough in the dataset to derive the factors. To get around this problem, the ratio of anomalous results to other results in the dataset needs to be increased. A simple way to accomplish this is to first select all observations in the dataset with the targeted result, then combine those observations with an equal number of randomly selected observations, thus yielding a 50/50 ratio.

- **Separation into training and validation datasets** – A common problem in data mining is that the output model of a data mining algorithm is **overfit** with respect to the **training data** – the data used to build the model. When this happens, the model appears to perform well when applied to the training data, but performs poorly when applied to a different set of data. When this happens we say that the model does not **generalize** well. To detect and assess the level of overfit or lack of generalizability, before a data mining algorithm is applied to a dataset, the data is randomly split into training data and **validation data**. The training data is used to build the model and the validation data is then applied to the newly built model to determine if the model generalizes to data not seen at the time of model construction.

- **Missing values** – Frequently, datasets are missing values for one or more attributes in an observation. The values may be missing because at the time the data was captured they were unknown or, for a given observation, the values do not exist.

  Since many data mining algorithms do not work well, if at all, when there are missing values in the dataset, it is important that they be handled before presentation to the algorithm. There are three generally deployed ways to deal with missing values:

  o   Eliminate all observations from the dataset containing missing values.

  o   Provide a default value for any attributes in which there may be missing values. The default value for example, may be the most frequently occurring value in an attribute of discrete types, or the average value for a numeric attribute.

  o   Estimate using other attribute values of the observation.

## Algorithm selection and application

Once the dataset has been properly prepared and an initial exploration has been completed, you are ready to apply a data mining algorithm to the dataset. The choice of which algorithm to apply depends on the objective of your data mining task and the types of data available. If the objective is classification, then you will want to choose one or more of the available classification modelers. If you are predicting numeric output, then you will choose from an available regression modeler.

Among modelers of a given type, you may not have a prior expectation as to which modeler will generate the best model. In that case, you may want to apply the data to multiple modelers, evaluate, then choose the model that performs best for the dataset.

At the time of model building you will need to have decided which attributes to use as input attributes and which, if building a prediction model, is the output attribute. (Cluster, association, and sequence analyses do not have an output attribute.) The choice of input attributes should be guided by relationships uncovered during the initial exploration.

Once you have selected your modelers and attributes, and taken all necessary steps to prepare the dataset, then apply that dataset to the modelers – let them do their number crunching.

## Model evaluation

After the modeler has finished its work and a model has been generated, evaluate that model. There are two tasks to be accomplished during this phase.

- **Model performance** – Evaluate how well the model performs. If it is a prediction model, how well does it predict? You can answer that question by either comparing the model's performance to the performance of a random guess, or by building multiple models and comparing the performance of each.

- **Model understanding** – Gain an understanding of how the model works. Again, if it is a prediction model, you should ask questions such as: "What input attributes contribute most to the prediction?" and "What is the nature of that contribution?" For some attributes you may find a direct relationship, while in others you may see an inverse relationship. Some of the relationships may be linear, while others are non-linear. In addition, the contributions of one input may vary depending on the level of a second input. This is referred to as variable interaction and is important to detect and understand.

## Summary

In this chapter an overview of a methodology for conducting a data mining analysis was presented. The methodology consists of four steps: initial data exploration, dataset preparation, data mining modeler application, and model evaluation. In the chapters that follow, readers will be guided through application of the methodology using a visual tool for data mining – VisMiner. Chapter 2 uses the visualizations and features of VisMiner to conduct the initial exploration and do some dataset preparation. Chapter 3 introduces additional features of VisMiner for dataset preparation not covered in Chapter 2. Chapters 4 through 7 introduce the data mining methodologies available in VisMiner, with tutorials covering their application and evaluation using VisMiner visualizations.

# 2

# Initial Data Exploration and Dataset Preparation Using VisMiner

## The Rationale for Visualizations

Studies over the past 30 years by cognitive scientists and computer graphics researchers have found two primary benefits of visualizations:

- potentially high **information density**

- **rapid extraction** of content due to parallel processing of an image by the human visual system.

Information density is usually defined as the number of values represented in a given area. Depending on the design, the density of visualizations can be orders of magnitude greater than textual presentations containing the same content.

In the vocabulary of cognitive science, a key component of rapid extraction of image content is usually referred to as **preattentive processing**. When an image is presented, the viewer's brain immediately begins extracting content from the image. In as little as 50 milliseconds it locates significant objects in the image and begins to categorize and prioritize those objects with respect to their importance to image comprehension. Researchers have identified a shortlist of visual properties that are preattentively processed – those that the brain considers to be of highest priority to which it initially directs its attention. They include: color, position, shape, motion, orientation, highlighting via

*Visual Data Mining: The VisMiner Approach*, First Edition. Russell K. Anderson.
© 2013 John Wiley & Sons, Ltd. Published 2013 by John Wiley & Sons, Ltd.

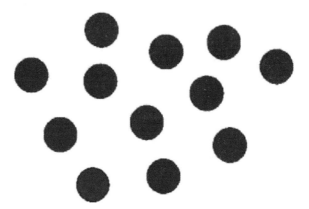

**Figure 2.1**   Color Preattentive Property

addition, alignment, and lighting anomalies. When visualizations are designed using these properties, attention can be immediately drawn to targets that the designer would like the viewer to focus on. Look at Figure 2.1, an image using color to preattentively focus attention. When you looked at the image, did the blue circle immediately "pop-out" or did you need to study the image for a short time before recognizing that the blue circle was different from the rest? Figure 2.2 illustrates the preattentive property of highlighting via addition.

In the design of visualizations, another aspect to consider is the capacity of human **working memory**. It is widely accepted that the human brain can actively hold in working memory a maximum of five or six items. Norman suggests that a good way to enhance human cognition is to create artifacts that externally supplement working memory. He refers to them as "things that make us smart". For example, something as simple as using pencil and paper to perform an arithmetic computation on multidigit numbers would be considered an external memory supplement. The visualizations of VisMiner were designed to be used as very effective external memory aids.

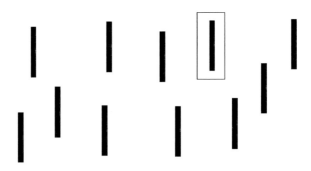

**Figure 2.2**   Addition Preattentive Property

Research suggests, however, that there are other issues to consider. Wolfe found that as an image is presented, the image is immediately abstracted, but details of the image are not retained in memory when focus shifts to a different image. According to Healey, "Wolfe's conclusion was that sustained attention to the objects tested in his experiments did not make visual search more efficient. In this scenario, methods that draw attention to areas of potential interest within a display [i.e., preattentive methods] would be critical in allowing viewers to rapidly and accurately explore their data".

Based on this research, VisMiner was created to present multiple, concurrent, non-overlapped visualizations designed specifically to be preattentively processed. The preattentive properties allow you to immediately recognize patterns in the data and the multiple, concurrent views supplement working memory as your eyes move from visualization to visualization when comparing different views of the dataset. As VisMiner was designed, information density was not considered as important when choosing the types of visualizations to be incorporated. Since information-dense visualizations require a relatively longer "study" time to extract patterns, they were not considered to be viable extensions to working memory.

## Tutorial – Using VisMiner

### Initializing VisMiner

If you have not already done so, start the VisMiner Control Center on one of the computers that you will be using for your data mining activities. Upon start-up, the Control Center will open in a maximized window shown in Figure 2.3. The Control Center is divided into three panes:

- **Available Displays** – The upper left pane depicts available displays that have been connected to the Control Center via VisSlave. Initially this pane is blank, because no slaves are currently connected.

- **Datasets and Models** – The lower left pane is used to represent open datasets, derived datasets, and models built using the open or derived datasets. Again, upon start-up, this pane is blank.

- **Modelers** – The right side pane contains icons representing the VisMiner data mining algorithms (or modelers) available for processing the datasets. Each is represented by a gear, and overlaid by the title of the modeler.

The Control Center interface is designed for direct manipulation, meaning that you perform data mining operations by clicking on or dragging and dropping the objects presented. Once you get started, you will find the use of VisMiner very intuitive and easy to learn.

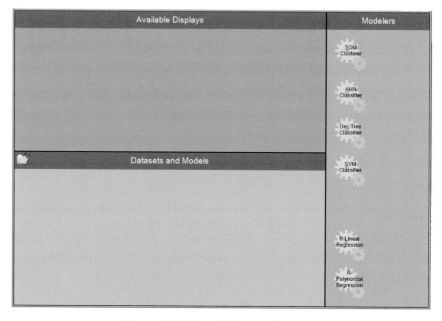

**Figure 2.3**   Control Center at Start-up

The Control Center is also designed to visually present the current status of your data mining session. All open datasets, derived datasets, models, and visualizations are represented as icons on screen. You should be able to quickly assess the current state of your activity by visually inspecting the Control Center icon layout.

## Initializing the slave computers

On each computer that you want to use to display visualizations, start the VisSlave component of VisMiner. If the same computer will be used for both the Control Center and the visualizations, then after starting the Control Center, also start VisSlave.

Upon start-up, VisSlave attempts to make a connection to the Control Center. If this is the first time that VisSlave has executed on the computer, it will prompt the user for the IP address of the computer where the Control Center is running. See Figure 2.4.

☞   Enter the Control Center's IP address.

☞   Select "OK"; a connection to the Control Center is established.

**Figure 2.4**   Control Center Prompt

On subsequent executions of VisSlave it will remember the IP address where it last made a successful connection and will attempt a connection without first prompting the user for an IP address. It will only prompt the user for the IP address, if it cannot find an active instance of the Control Center on the computer where it made its last successful connection. (If you do not know the IP address of the Control Center computer, see Appendix C for instructions.)

As each slave is started and a connection is made to the Control Center, the slave will report to the Control Center, the properties of all displays it has available. The Control Center will then immediately represent those displays in the "Available Displays" pane. See Figure 2.5 for an example of a slave

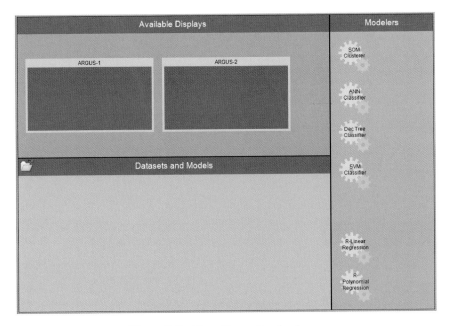

**Figure 2.5**   Slave Connection by Argus

computer named "Argus" that has made the connection. It reported two available displays.

After completing a VisMiner session, close the Control Center, which will notify all connected slaves and visualizations that they too need to shut down.

## Opening a dataset

VisMiner is designed to open datasets saved in comma-delimited text files (.csv) and in Microsoft Access tables (MDB files). If your data is not in one of these formats, there are many programs and "Save as" options of programs that will quickly convert your data. Let's begin by opening the file Iris.csv, which is contained in the data packet accompanying VisMiner. To open:

⏵ Click on the "File open" icon located on the bar of the "Datasets and Models" pane.

⏵ Complete the "Open File" dialog in the same way that you would for other Windows applications by locating the Iris.csv file. Note: If you do not see any .csv files in the folder where you expect them to be located, then you probably need to change the file type option in the "Open File" dialog.

## Viewing summary statistics

All currently open datasets are depicted by the file icon in the "Datasets and Models" pane. Start your initial exploration by reviewing the summary information of the Iris dataset. To see summary information:

⏵ Right-click on the Iris dataset icon.

⏵ Select "View Summary Statistics" from the context menu that opens.

The summary for the Iris dataset (Figure 2.6) gives us an overview of its contents. In the summary, we see that there are 150 rows (observations) in the dataset, and five columns (attributes). Four of the five attributes are numeric: PetalLength, PetalWidth, SepalLength, and SepalWidth. There is just one nominal attribute: Variety. For each numeric attribute, the summary reports the range (minimum and maximum values), the mean, and the standard deviation. Nominal attributes have cardinality. The cardinality of Variety is 3, meaning that there are three unique values in the dataset. You can see what those values are by hovering over the cell in the Cardinality column at the Variety row. As you hover, the three values listed are "Setosa", "Versicolor", and "Virginica". The number in parentheses following the value indicates how many observations there are containing that value. In the Iris dataset there are 50 observations of variety Setosa, 50 of Versicolor, and 50 of Virginica.

Dataset: Iris.csv

Rows: 150

| Column Name | Type | Minimum | Maximum | Mean | Std Deviation | Cardinality |
|---|---|---|---|---|---|---|
| PetalLength | Real | 1.00 | 6.90 | 3.76 | 1.759 | N/A |
| PetalWidth | Real | 0.100 | 2.500 | 1.199 | 0.761 | N/A |
| SepalLength | Real | 4.30 | 7.90 | 5.84 | 0.825 | N/A |
| SepalWidth | Real | 2.000 | 4.400 | 3.054 | 0.432 | N/A |
| Variety | Text | N/A | N/A | N/A | N/A | 3 |

**Figure 2.6**   Iris Summary Statistics

You can sort the rows in the summary by clicking on a column header. For example, to sort by mean:

☞   Click on the "Mean" column header.

☞   Click a second time to reverse the sort.

☞   Select "Close" when you have finished viewing the summary statistics.

You have now completed the first two tasks in the "initial data exploration" phase – determining the dataset format and attribute identification.

## Exercise 2.1

The dataset OliveOil.csv contains measurements of different acid levels taken from olive oil samples at various locations in Italy. This dataset, in later chapters, will be used to build a classification model predicting its source location given the acid measurements. Use the VisMiner summary statistics to answer the questions below.

a. How many rows are there in the dataset?

b. List the names of the eight acid measure attributes (columns) contained in the dataset.

c. How are locations identified?

d. Which acid measure has the largest mean value?

e. Which acid measure has the largest standard deviation?

f. List the regions in Italy where the samples are taken from. How many observations were taken from each region?

g. List the areas in Italy where the samples are taken from. How many observations were taken from each area?

## The correlation matrix

After viewing the summary statistics for the Iris dataset, evaluate the relationships between attributes in the data. In VisMiner, a good starting point is the correlation matrix.

To open the Iris dataset in a correlation matrix viewer, drag the dataset icon up to an available display and drop. A context menu will open, listing all of the available viewers for the dataset.

Select "Correlation Matrix".

The correlation matrix (Figure 2.7) visually presents the degree of correlation between each possible pairing of attributes in the dataset. Direct correlations are represented as a shade of blue. The more saturated the blue the stronger the correlation. Inverse correlations are represented as a shade of red, again with saturation indicating the strength of the correlation. Between pairings of numeric attributes, the coefficient of correlation is encoded using the blue or red colors. Between pairings of nominal and numeric attributes the

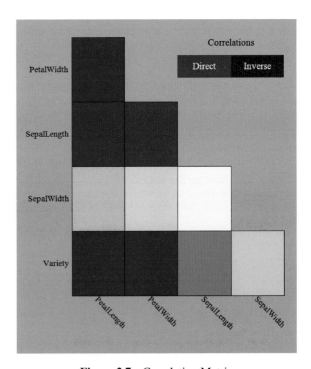

**Figure 2.7**   Correlation Matrix

**eta coefficient** is used. Eta coefficients range in value between 0 and 1. There is no inverse relationship defined for correlations between numeric and nominal data types. Between pairings of nominal attributes, the **Cramer coefficient** is computed. Like the eta coefficient, it too ranges in value between 0 and 1, since there is no such thing as an inverse relationship.

The saturated colors support preattentive processing. A quick glance at the matrix is all that is needed to identify highly correlated attributes.

⮕   When you need a more precise measure of correlation, use the mouse to hover over a cell. As you do so, the actual correlation value is displayed within the cell.

The correlation matrix also has a feature to support the dataset preparation step – specifically dimension reduction. To remove an attribute from the matrix, simply ctrl-click on the attribute name along the side or bottom of the matrix.

⮕   For example, if you wanted to create a dataset containing only the numeric attributes, ctrl-click on the Variety label. Immediately that attribute is removed from the matrix.

As you exclude attributes, a list appears to the upper-right of the matrix showing which attributes have been removed. If you remove an attribute by mistake or later change your mind, you can click on the attribute in the list to restore it to the matrix. (See Figure 2.8.)

Whenever there are excluded attributes, another button ("Create Subset") appears to the left of the list.

⮕   To create a dataset without the eliminated attributes, select the "Create Subset" button.

You will be prompted to enter a name for the new dataset and an optional description.

⮕   Enter the name "Iris Measures".

⮕   Select OK.

The Correlation Matrix notifies the Control Center that you are creating a derived dataset from the Iris dataset with details on its contents. The Control Center creates this dataset and represents it in the Datasets and Models pane as a descendent of the base Iris dataset (Figure 2.9). Derived sets, such as the one just created, exist only within VisMiner. To save for use in a later VisMiner session, right-click the derived set, then select "Save as dataset".

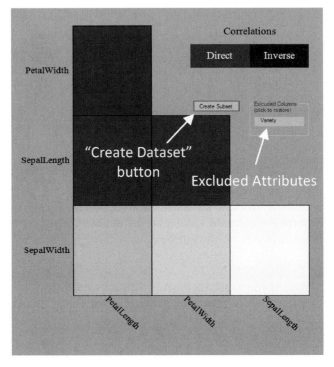

**Figure 2.8**   Correlation Matrix with Variety Excluded

**Figure 2.9**   Derived Dataset from Correlation Matrix

## Exercise 2.2

Use the VisMiner correlation matrix to answer the questions and perform the operation below with respect to the OliveOil dataset.

a.  List the three strongest inverse correlations between acid levels. (Do not include correlations with the location attributes AreaName and Region-Name.) What is the coefficient of correlation of each?

b. List the three strongest direct correlations between acid levels. What is the coefficient of correlation of each?

c. Which acid is most strongly correlated with Region? What is the eta coefficient?

d. Which acid is most strongly correlated with Area? What is the eta coefficient?

e. Create a derived set of the OliveOil data that contains acid measures only. Name the derived set "Acids".

## The histogram

A histogram visually represents value distributions of a single attribute or column.

↶⊃ Drag the Iris dataset up to an available display and drop. A context menu will open listing all of the available viewers for the dataset.

↶⊃ Select "Histogram".

The distribution of the Variety column is shown in Figure 2.10a. By default, Variety is the first column selected by the histogram. It looks rather boring. There are three bars – one for each of the three varieties in the dataset. The bars are all of equal height; because in the dataset there are 50 observations for each variety.

↶⊃ Using the "Column" drop-down, change the column selection to "PetalLength".

The PetalLength distribution is a little more interesting (Figure 2.10b). Notice the gap between bars in the 2–3 centimeter range. Very clearly we see a multimodal distribution. The observations on the left do not appear to have been drawn from the same population as those on the right.

The histogram bars are defined by first the column value range into a predetermined number of equal sized buckets. In the VisMiner histograms when numeric data is represented, by default VisMiner chooses 60 buckets. Once the number of buckets is determined, each observation is assigned to the bucket corresponding to its value, and the number of observations in the bucket is encoded as the height of the bar. The bucket containing the most observations is drawn full height and the others, based on their observation counts, are sized relative to the tallest.

At times, depending on the range of each bucket, the highs and lows of neighboring bars will vary significantly based on the chosen bucket count. Slight adjustments in the bucket range can produce large changes in the heights

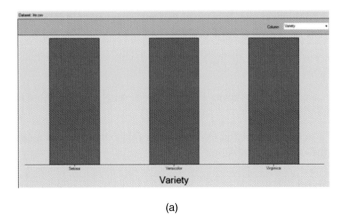

(a)

(b)

**Figure 2.10**   Histogram

of neighboring bars. The overcome this, the process of smoothing adjusts each bar's height by also factoring the height of neighbors.

☞   Click on the "+" button in the title bar to the right of the "Smooth" label to somewhat smooth the bar heights.

Notice the adjustments in bar height.

☞   Repeatedly click on the "+" button until it disappears.

Each time the "+" button is clicked, the neighboring bars contribute more and more. At the highest possible level of smoothing, we see two and maybe three sub-populations of somewhat normally distributed values. During initial exploration, when we see multimodal distributions similar to the PetalLength distribution, it is a strong indicator that in subsequent exploration and algorithm application, we should attempt to understand why. (Note: Smoothing of a distribution can be decreased and returned to its unsmoothed level by repeatedly clicking I the "−" button.

## The scatter plot

A third available viewer in exploring a dataset is the scatter plot – useful for evaluating the nature of relationships between attributes. The scatter plot is probably a familiar plot to you as it represents observations as points on X-Y (and Z) axes.

To open a scatter plot of the Iris dataset, drag the dataset up to the icon representing the display currently being used for the correlation matrix. As you drag the dataset over the display icon, a dashed rectangle is drawn showing you where the new plot will be created. As you drag to the left, the rectangle will be on the left pushing the current correlation matrix to the right side of the display (Figure 2.11). As you drag to the right, you will first see the rectangle fill the entire display, indicating that the new plot will replace the correlation matrix. Continuing to the right, the rectangle moves to the right side of the display pushing the correlation matrix to the left. Drop the dataset at this location.

Select "Scatter Plot" from the context menu. The plot is opened and the correlation matrix is pushed to the left side of the display (Figure 2.12).

By default the scatter plot shows the first two attributes in the dataset on the X and Y axes respectively. Notice that there is no scale on either of the axes. This is intentional. In VisMiner, scatter plots are intended to represent relationships between attributes, not to be used for point reading. The only indicator of scale

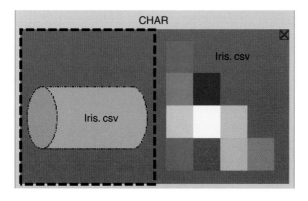

**Figure 2.11**   Viewer Location

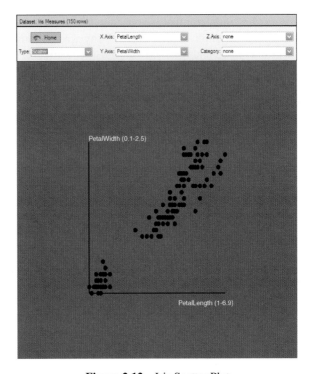

**Figure 2.12**    Iris Scatter Plot

is found inside the parentheses following the attribute name, where the minimum and maximum values (range) are shown. To get a point reading, hover over one of the axes at a desired location.

☞    Hover over the X axis at about its midpoint. What is the value for PetalLength at that location?

Looking back at the Control Center you will also notice an arrow pointing from the correlation matrix to the scatter plot. This indicates that, because they represent the same dataset, the scatter plot may be manipulated using the correlation matrix.

☞    Try this out by clicking on the almost white cell in the correlation matrix representing SepalLength and SepalWidth. Immediately, the scatter plot changes its axes to the two selected attributes of the correlation matrix.

This feature allows you to quickly browse plots of any attribute combinations that correspond to cell pairings in the correlation matrix.

When looking at scatter plots of the data, shortcomings of the correlation matrix become apparent. Correlations are one very simple measure of relationships between attributes; yet they hide detail. For example, in the correlation matrix for the Iris data, you see a relatively strong inverse relationship between SepalWidth and PetalLength. (Coefficient of correlation is −0.421.) Indicating that as PetalLength increases, SepalWidth decreases. This is somewhat counterintuitive. One would expect that as the size of the flower increases all measures would increase.

☞ To evaluate this relationship, click on the SepalWidth/PetalLength cell in the correlation matrix.

Look at the resulting scatter plot (Figure 2.13). Notice the two clusters of plot points – one below and to the right of the other – which resulted in the inverse correlation.

☞ Continue your inspection by selecting Variety in the "Category" dropdown in the options panel above the plot.

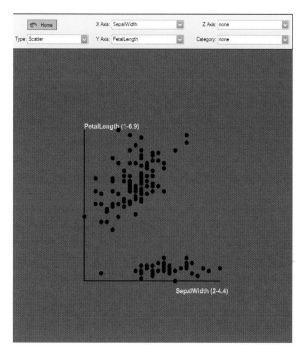

**Figure 2.13**   Sepal Width versus Petal Length

Each Variety in the dataset is now represented using a different color. You should recognize that the clusters of points represent different varieties. Setosa is in the lower right cluster. Versicolor and Virginica are in the upper left. You should also note that within the Versicolor-Virginica cluster there is a direct relationship between PetalLength and SepalWidth rather than the inverse relationship reported by the correlation matrix.

Suppose that the objective of your data mining activity is to determine a set of classification rules to predict iris variety based on the four flower measures. The scatter plot can help you formulate those rules. For example, in the plot of PetalLength versus PetalWidth, with Variety selected as the category (Figure 2.14), you clearly see that Setosa flowers are much smaller. You also see that Versicolor are next in size; Virginica are the largest. Note also that although there is a distinct separation between Setosa and the others there is some overlap between the Versicolor and Virginica. It will be more difficult to distinguish between these two varieties.

You can add a third (Z) dimension to the scatter plot by selecting another attribute using the "Z Axis" drop-down. Try selecting SepalWidth. Static 3-D

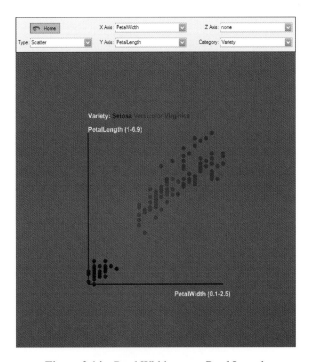

**Figure 2.14**   Petal Width versus Petal Length

plots drawn on a 2-D display can be difficult to interpret and see accurate point positioning.

To assist in this task, the scatter plot (and other 3-D plots in VisMiner) allows you to rotate the plot by clicking and dragging in the direction that you would like to rotate. Using the rotation, your brain can preattentively and accurately evaluate the 3-D positions of points in the plot and assess relationships in 3-D space.

The "Home" button returns the plot to its original position.

Explore another feature of the scatter plot by:

selecting PetalWidth for the X axis, "none" for the Z axis and Variety for the Y axis. As you select Variety, a nominal attribute, the type of the plot switches from a true scatter plot to a **height density** or histogram (Figure 2.15). Again you can rotate the plot to get a better feel for the histogram heights.

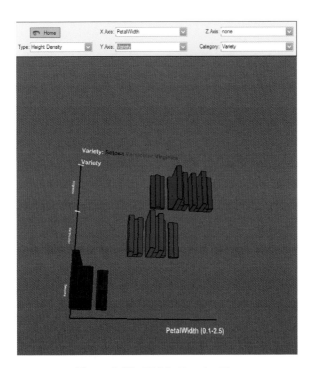

**Figure 2.15**   Height Density Plot

The Scatter Plot viewer automatically adjusts the plot type depending on the type of attributes selected and the number of observations in the dataset. You will see more plot types in later tutorials as you explore additional datasets.

⊄   Using the Control Center, close the correlation matrix and the scatter plot, by clicking on the small red "X" in the upper right corner of each viewer.

Exercise 2.3

Use the VisMiner correlation matrix, histogram and scatter plot to answer the questions below with respect to the OliveOil dataset.

The correlation matrix reports an inverse correlation between eicosenoic and oleic acids ($-0.424$). Does it fully reflect the actual relationship between these two acids? Explain your answer using snapshots of any scatter plots that you used to answer this question.

a. Evaluate the distribution of stearic, eicosenoic, and palmitic acid measurements using the histogram viewer. Do they appear close to normal, skewed, or multimodal?

b. Develop two rules (guidelines) that could be used to classify observations into regions. For example, a rule might be: If x > 50, then assign to category A. Be sure to include at least one rule that differentiates between the North and Sardinia regions. Include selected scatter plots to support your rules.

The parallel coordinate plot

The parallel coordinate plot (PCP) is the most powerful of the available visualizations in VisMiner and due to that power can take more time to interpret. Not all of the patterns are preattentive.

⊄   To open, drag the Iris dataset up to the display and release, then select "Parallel Plot" in the context menu (Figure 2.16).

One of the strengths of the PCP is that it is not limited in the number of dimensions that it can concurrently plot. As the name implies, each dimension is plotted along axes that are laid out in parallel rather than orthogonal axes like a scatter plot. Each observation is represented as a sequence of line segments connecting points on each of the axes. To help you understand the PCP, let's focus on just one observation. There are small red triangles at the top and bottom of each of the axes. The triangles function as range end-points or **filters**. Only

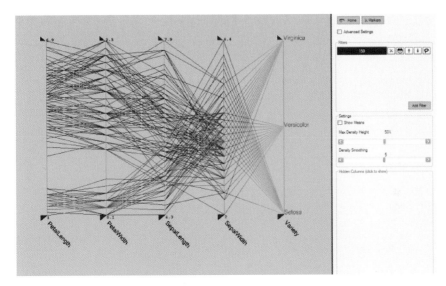

**Figure 2.16**   Parallel Coordinate Plot

observations between the top and bottom filters are drawn. Initially all observations are included given that the triangles are all the way to the top and bottom of their respective axes.

To filter out all but a single observation, drag the triangle at the top of the SepalWidth axis down to near the bottom. As you drag, observations outside the range of the filter are removed from the plot. Keep dragging until just one observation remains (Figure 2.17).

As with the scatter plot, scales for the axes are intentionally omitted. You can read point values along the axes by hovering over locations up and down an axis.

To read values for the depicted observation, first hover over the intersection of the PetalLength axis and the line segment for the observation. As you hover, the value 3.5 appears next to the mouse pointer, meaning that for this observation, its PetalLength is 3.5 centimeters. Next hover over the intersection of the line segment with the PetalWidth axis to see a value of 1, then over SepalLength to see a value of 0.5, and on to SepalWidth for a value of 2. Finally, you see the line segment end at Versicolor, telling you the Variety of this observation.

Drag the filter triangle back to the top of the SepalWidth triangle to redisplay all of the observations.

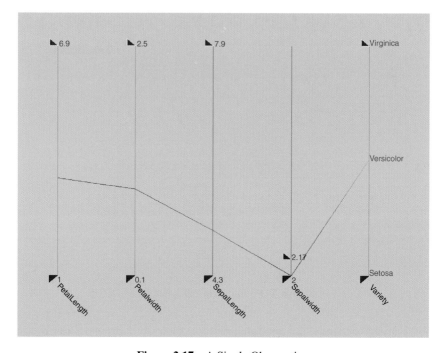

**Figure 2.17**   A Single Observation

Look at the overall plot. You see a cluster of observations with short and narrow petals. Based on your experience in viewing the same data in the Scatter Plot, can you guess which variety these observations belong to?

To verify, drag the top triangle of the Variety axis down to a location between Versicolor and Setosa. Once you pass Versicolor, only the Setosa are visible (Figure 2.18).

The overall interpretation that you can make from the PCP at this arrangement is that Setosa iris flowers have relatively short and narrow petals, somewhat short sepals, but wider sepals relative to the other iris in the dataset.

Let's now compare the three varieties.

Click on the "Add Filter" button.

A blue filter is added. It includes all Iris data observations, since the blue filter triangles are at the top and bottom of each axis. You don't see all of the blue triangles because they are behind the red triangles. A fast way to concurrently

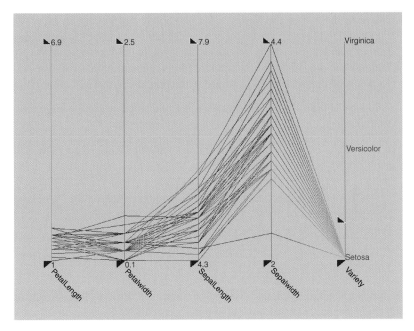

**Figure 2.18**    Filtered for Setosa Only

slide multiple triangles is to use the lasso located at the right end of the filter box (Figure 2.19) in the options panel on the right side of the display.

Click on the lasso at the end of the blue filter bar. The cursor will change to a cross-hair indicating that you are in lasso mode.

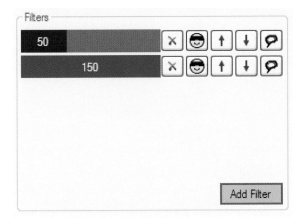

**Figure 2.19**    Filter Box

⇨ Move to the left of the Variety axis just above the Versicolor label. Drag down and to the right across the Versicolor label. As you drag a rectangular rubber band is drawn representing the area to be included in the filter. Release the drag and the blue triangle filters are automatically moved to the location of the rubber band where it intersects each axis.

⇨ Add a third filter for the Virginica variety and select using the lasso tool for that filter.

You now see a complete picture of the iris varieties using color. The Setosa are in red, the Versicolor in blue, and the Virginica in green (Figure 2.20).

⇨ To individually inspect each of the filter groups, click on the "hide/show" icon in the filter box for the other two filters. For example, hide the red and blue filters to leave just the green (Virginica).

⇨ Compare the Virginica observations with the Versicolor by unhiding the blue (Versicolor) filter.

At times there are so many observations that it is difficult to compare filter groups because the observations of the top obscure those underneath. When this happens, an alternative option is to compare the filter means only.

⇨ In the "Settings" box of the options panel on the right, check the "Show Means" box.

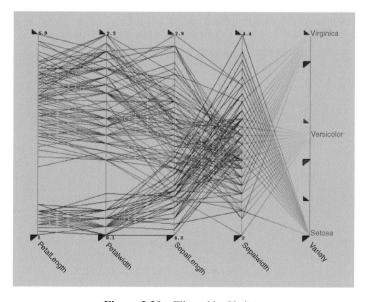

**Figure 2.20**   Filtered by Variety

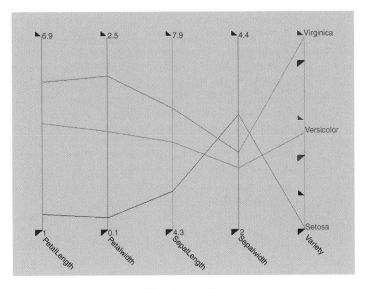

**Figure 2.21**   Filter Means Only

Now instead of one line sequence for each observation in the filter group, you see one line sequence representing the mean values for each filter group and can readily compare one group to the other (Figure 2.21).

## Exercise 2.4

Use the VisMiner parallel coordinate plot to answer the questions below with respect to the OliveOil dataset. To keep the region and area names from being obscured by the plot lines, it is best to drag the AreaName axis all the way to the right and to drag the RegionName axis next to it on the left. Flatten both axes.

a. Create separate filters for each of the three regions. Which areas belong to the South region; which areas belong to the North region; which areas belong to the Sardinia region?

b. Use the PCP to develop classification rules to assign observations to a Region similar to what you did in exercise 2.3c. Hint: Open a second PCP with all observations belonging to the default initial filter group. Using the first PCP (with the three separate filter groups) as a guide, adjust the acid sliders in the second PCP to eliminate all observations belonging to the South region without eliminating any belonging to the other two. You can tell if you have accomplished this when there are no line segments connecting to the South region and the number of remaining observations is 249 (the number in the Sardinia and North regions combined). Note the position of any sliders that you adjusted to eliminate the South region observations. The positions of these sliders become a classification rule. Now hide the South region filter group in the first PCP in order to focus on the other two. Again using the first PCP as a guide, adjust the sliders in the second PCP to eliminate all observations of one region while leaving all observations of the other region visible. Once complete, the positions of these sliders become the second classification rule.

c. Working with the first PCP, the one with the three filter groups, hide the North and South region filter groups leaving only the Sardinia visible. Looking at the linoleic axis, you see two distinct sub-populations. How are these sub-populations related to area?

To explore the data distribution assessment features of the PCP:

↪ Close the PCP for the Iris data.

↪ Open the file Pollen.csv.

↪ View its "Summary Statistics".

The Pollen dataset contains five measures of grains of pollen: Crack, Density, Nub, Ridge, and Weight. There are almost four thousand observations. The dataset is actually a synthetic dataset created for use in a data mining competition. The objective of the competition was to find significant sub-populations within the dataset. We will use it to assess data distributions.

↪ View the Pollen data in a PCP by dragging its dataset icon up to a display and selecting "Parallel Plot".

The initial PCP of the Pollen data (Figure 2.22) looks quite different from that of the Iris data due to its much larger observation count (3838 compared to 150). Only a few of the individual observation line segments are distinguishable. To assist in evaluating the distributions of each attribute, the densities are color encoded. The lighter shades of red, transitioning toward yellow, indicate areas of greater observation density. In VisMiner, the PCP is actually a 3-D plot. To view:

↪ Rotate the plot to the right or left by dragging, and you will see that the areas of greater density, in addition to being color encoded are also raised up from the surface of the plot. By rotating about 20 degrees to the left, you can clearly see the distributions of the Crack attribute at the left end and the Weight attribute at the right end.

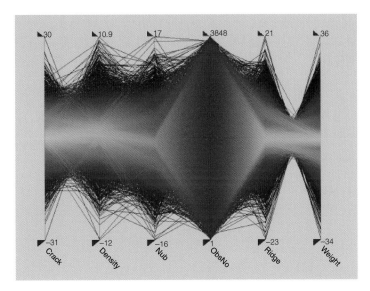

**Figure 2.22**   PCP Pollen Dataset

Both attributes appear to be somewhat normally distributed, yet both have some irregularly shaped (not smooth) peaks. When rotating to the right or left it is easy to see the distributions of the attributes on either end, but not so easy to see distributions of attributes in the middle. To clearly see the distributions of attributes currently displayed on the middle axes you would need to move them to the outside.

To do this, hover over the axis of the attribute that you wish to reposition, then drag it to the right or left. You can tell when you are hovered over an axis when you see the attribute name enlarge and change from white to yellow.

Move the ObsNo attribute all the way to the right.

Look at its distribution. What do you see? The distribution is flat because observation numbers are a sequential numbering starting with 1 and going to 3,848. Obviously, this attribute should not be included in any data mining operation, since it should have no relationship with any of the other attributes.

Eliminate ObsNo from the plot by again hovering over the axis and right-clicking. Select "Hide ObsNo" from the context menu.

You should now see a plot similar to Figure 2.23. Notice the difference in pattern between line segments connecting Weight and Ridge versus the patterns

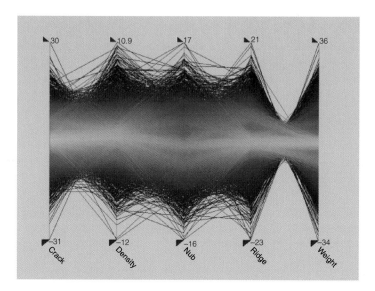

**Figure 2.23**  PCP ObsNo Removed

between other adjacent attributes. The crossing pattern that you see between Weight and Ridge indicates an inverse relationship. Low values for Ridge correspond to high values for Weight and vice versa. Direct relationships would be seen as line segments that are close to parallel. There are no strong direct relationships in this dataset.

One shortcoming of parallel plots is that you only visually see the relationships between adjacent attributes. To see all of the relationships, you would need to systematically drag each of the axes to be adjacent to each of the other axes, pausing as you drag to observe the connection pattern between the axes. For example, suppose your current axis sequence is: Crack, Density, Nub, Ridge, and Weight and you would like to see all the relationships between Crack and the other attributes.

- Drag Crack to the right until it is situated between Density and Nub. Pause to assess the relationship between Crack and Density to the left of the Crack axis and Crack and Nub on the right. Next drag *Crack* to the right again until it is situated between Ridge and Weight. Pause again to assess the relationships.

- Now rotate the plot all the way around (180 degrees) to look at the back side (Figure 2.24). When you see a pattern similar to the patterns between Density, Nub and Ridge they are indicative of independence (lack of correlation) between the attributes.

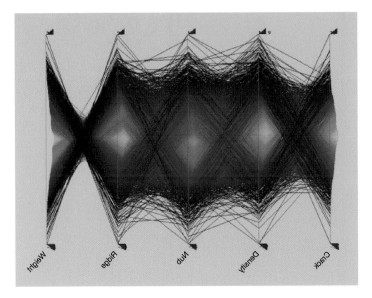

**Figure 2.24**   PCP Backside

If desired, you could confirm this independence by loading the Pollen data into a correlation matrix.

### Extracting sub-populations using the parallel coordinate plot

The PCP is an excellent tool to use when visually searching for sub-populations within a dataset. A good indicator that there are sub-populations within a dataset can be observed by looking for ribbons of lighter shading in the plot. Do you see any in the Pollen data of Figure 2.23?

Another indicator is to look at the attribute distributions by rotating slightly to the right or left in order to see the densities based on height. When you see a multimodal (multiple peaks) distribution, it is likely that each peak represents a sub-population present in the full dataset (Figure 2.25).

☞ Drag the Ridge axis to the right end.

☞ Rotate slightly to see two very distinct peaks on that Ridge axis. The tallest is at about mid-level on the axis where you would expect the peak of a normal distribution to be located, although the peak is higher than would be expected of a normal curve. The second peak is about two-thirds of the way up from the bottom of the axis.

To explore these sub-populations in detail, we need to extract them from the full dataset. We'll do the tallest peak first – the light colored ribbon that we see

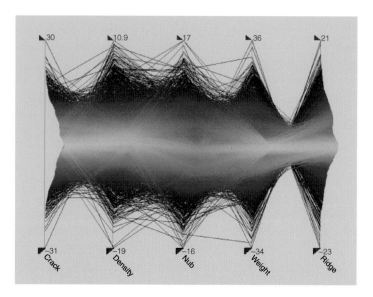

**Figure 2.25**    PCP with Multi-Modal Peaks at Right

running near the centers of each axis. We start the process by using the lasso tool to quickly narrow down all axis filters to the target ribbon. See Figure 2.25 for a view of the initial state before any filter sliders have been adjusted.

Click on the lasso tool, then move all the way to the left of the Crack axis at a height safely above the target ribbon.

Drag the rubber band to the right and down across all axes to a point safely below the ribbon on the Ridge axis.

Release at this point.

The plot should look similar to Figure 2.26. Now it is time to fine-tune the filters. Begin with the Weight filters since the target ribbon on that axis is quite visible.

Slide the top and bottom filter triangles toward the middle to eliminate most of the observations except for those within the target ribbon. As you narrow the Weight filters down, the ribbon becomes more distinctive near the four other axes.

As needed, slide each of those filter triangles closer to the ribbon until you have eliminated all non-ribbon observations. Your end result should look similar to Figure 2.27.

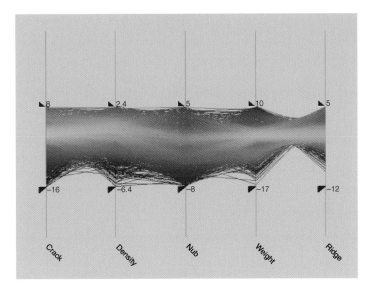

**Figure 2.26** PCP after Lasso Selection

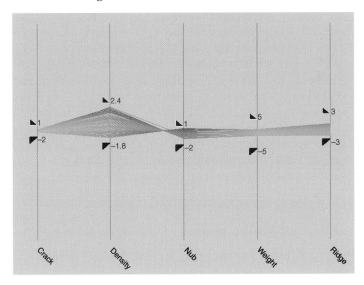

**Figure 2.27** PCP with Isolated Sub-population

Once the target ribbon has been isolated, you are ready to extract those visible observations to a new dataset.

Right-click on any of the filter triangles, then select "Make dataset from filter".

↺  You will be prompted for a name and description. At a minimum, give it a name.

↺  Select "Create".

The new dataset now appears on the Control Center – derived from the Pollen dataset.

↺  View the summary statistics for this subset. Notice that, using the PCP, we have accomplished both observation reduction (almost 4,000 observations down to about 100) and dimension reduction (ObsNo was removed from the plot).

↺  Close the summary statistics.

↺  View the derived set in a scatter plot.

What do you see? Has your data mining activity found any patterns of interest?

There are numerous other options in the PCP that have not yet been described. You are encouraged to experiment with them on your own. Many of the options become available with a right-click of either a range slider for filter options, or an axis. All features of the filter option are also available using the icon buttons in the options panel on each filter bar, except for the "Make Dataset . . . " option.

You have already used the "Hide" option of an axis to remove it from the plot. The "Add Marker" option assists in point value reading. As noted earlier, the axes were designed without scales in order to reduce chart clutter and to get the user to focus more on patterns than value reading. If you need to know the value at a given location on an axis, then hover over that location with the mouse pointer and the value at the hover location will be displayed. If you would like the value to remain after moving the mouse pointer away:

↺  Right-click at the desired location on the axis.

↺  Select the "Add Marker" option.

↺  Move to a different location along an axis, then add another marker.

↺  Hover over an existing marker and right-click to remove the marker or alternatively . . .

↺  Select the delete "Markers" button in the options panel to remove all displayed markers.

When plotting density heights on the 3-D Z axis, VisMiner uses the point of greatest density over all axes as the top point. That top point is always plotted at a predefined height. All other density heights are plotted relative to the maximum. The "Flatten" axis option is useful when you have an axis with a dominant distribution value. For example, you may have a Yes/No nominal attribute in which 90% of the observations are "Yes". The density height at the "Yes" location on the axis will be so much greater than all other density points that it totally dominates and makes distribution assessments along the other axes difficult. To prevent that domination, right-click on the axis that dominates and select the "Flatten" option. That axis will be drawn with density height encoding and the other axes will adjust their heights relative to the next best height density point. It sounds complicated, but remember, if you see a single point dominating all others with respect to height density, making the others look flat, then flatten that axis and the others will be allowed to grow.

Exercise 2.5

Use the VisMiner parallel coordinate plot to locate and extract sub-populations from the out5d.csv dataset. This dataset contains satellite and sensor readings collected from a 128 × 128 grid in Western Australia – SPOT a measurement from a satellite image, magnetics, and three radiometric measures: potassium, thorium, and uranium. The X and Y attributes represent the grid column and row respectively of each observation's measurement. After opening the dataset in the PCP, hide the X, Y, and ObsNo axes.

a.  Inspect the distribution of observations about the Uranium axis. How many sub-populations (peaks) are seen?

b.  Adjust the sliders on the Uranium axis to isolate the upper sub-population. Within that newly isolated sub-population, how many distinct sub-populations are visible looking at the either the Magnetics or Potassium axes? Are the visible sub-populations of the Potassium axis the same as those of the Magnetics axis? Explain your answer.

c.  Adjust the sliders on the Magnetics axis to isolate the lower sub-population. What is the relationship between Magnetics and Potassium?

d.  Make a dataset of the isolated sub-population, giving it a name of "LowMagnetics". Be sure that the dataset contains all attributes including ObsNo, X, and Y. Save the newly created derived set to disk to use in a later exercise.

To summarize, as the parallel plot was introduced, it was described as the most powerful of the VisMiner visualizations and, as a result, the most complex in its use. In the preceding examples, it was used to locate patterns between attributes, to assess data distributions, to visually compare sub-populations, and to isolate and extract sub-populations. As you look back, many of the tasks introduced in Chapter 1 under the initial exploration and dataset preparation steps, may be completed using the parallel plot. In later examples you will use the parallel plot for:

- outlier detection and removal

- more sophisticated dataset preparation tasks

- comparing clusters generated by a cluster analysis.

The table viewer

The table viewer is not an actual visualization, but a simple tabular layout of dataset values. It is included in VisMiner as a supplement to the other visualizations to be deployed when the user needs to see actual values.

☞   Drag the Iris dataset up to a display.

☞   Select "Table View".

The dataset opens in a table containing all values in the dataset (Figure 2.28). Its features are simple. Click on a column (attribute) heading to sort the observations by that attribute. Click again to reverse the sort sequence. Additional features will be introduced in later examples as the table data is synchronized with other viewers.

| PetalLength | PetalWidth | SepalLength | SepalWidth | Variety |
|---|---|---|---|---|
| 1.5 | 0.2 | 5 | 3.4 | Setosa |
| 3.8 | 1.1 | 5.5 | 2.4 | Versicolor |
| 1.5 | 0.4 | 5.7 | 4.4 | Setosa |
| 1.3 | 0.2 | 4.4 | 3.2 | Setosa |
| 5.3 | 1.9 | 6.4 | 2.7 | Virginica |
| 4.5 | 1.5 | 5.6 | 3 | Versicolor |
| 5.6 | 1.4 | 6.1 | 2.6 | Virginica |
| 1.4 | 0.2 | 5 | 3.6 | Setosa |
| 4.1 | 1 | 5.8 | 2.7 | Versicolor |
| 6.7 | 2 | 7.7 | 2.8 | Virginica |
| 5.5 | 1.8 | 6.4 | 3.1 | Virginica |
| 1.6 | 0.2 | 5 | 3 | Setosa |
| 1 | 0.2 | 4.6 | 3.6 | Setosa |
| 1.3 | 0.3 | 4.5 | 2.3 | Setosa |
| 3.7 | 1 | 5.5 | 2.4 | Versicolor |
| 5.8 | 1.6 | 7.2 | 3 | Virginica |
| 4.4 | 1.3 | 6.3 | 2.3 | Versicolor |
| 4.2 | 1.5 | 5.9 | 3 | Versicolor |
| 1.3 | 0.2 | 4.4 | 3 | Setosa |

**Figure 2.28**   Table View – Iris Dataset

The boundary data viewer

The boundary data viewer is a specialized viewer for comparing data measures by geographic or political boundaries. Current boundaries supported are US states, US congressional districts, US counties, three-digit US zip codes, and five-digit US zip codes. The viewer also supports datasets containing both a boundary component and a temporal component. For example, you may have a dataset containing population by state (the boundary component) and by year (the temporal component). If your dataset does not contain a temporal component, then in order to use this viewer, a boundary identifier (state, county, etc.) must be present, and it must uniquely identify observations in the dataset. For example, if you want to compare data by state, then there must be an attribute in the dataset to identify the state (state name, state abbreviation, or state FIPS code) and there must be just one observation per state in the dataset. If your data also contains a temporal component, then each observation in the dataset must be uniquely identified by a combination of the boundary value and the temporal value. For example, if you want to compare data by state from year to year, then there must be an attribute to identify the state and a second to identify the year. The combination of state and year must uniquely identify an observation.

↪ Open the file DataByState.csv.

↪ View the summary statistics for the dataset.

Notice that there are 52 rows in the dataset (it includes Washington DC and Puerto Rico). The attribute named State contains the state name. There are six other numeric data measures in the dataset that have abbreviated names. For example, the attribute PcntCollegeGradOvr25 is the percentage of the population over 25 years of age that are college graduates.

↪ Close the summary statistics.

↪ To start the viewer, drag the dataset to a display, and select "Boundary Plot".

Before opening the actual viewer, you will be prompted, in a pop-up dialog, to specify some parameter values that the viewer will need to organize the data. See Figure 2.29.

↪ The first requested parameter is the boundary for which your dataset contains its data. For this dataset choose "State".

↪ The second parameter requested is the name of the attribute in the dataset containing the boundary identifier. In this dataset choose the attribute named State.

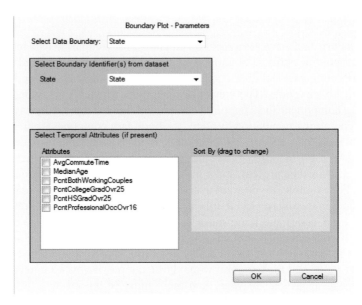

**Figure 2.29**   Boundary Plot Parameters

☞  Third, if the data contains a temporal component, you are asked to specify which attributes contain the temporal values. Since this data has no temporal component, leave the selection blank.

☞  Select "OK".

When the viewer initially opens, there is no data selected to show.

☞  Before selecting the data to show, zoom in on just the 48 contiguous states by stretching a rubber band over those states. Click and drag from above and to the left of Washington past Maine on the right and down to fully include Florida on the bottom, then release.

☞  In the "Show data for" drop-down, select MedianAge. (See Figure 2.29.)

The data is encoded using color similar to the encoding in the correlation matrix (Figure 2.30). A pale yellow represents the low value, while a fully saturated blue represents the high value. The color encoding enables a pre-attentive pop-out effect. Without conscious analysis, you should immediately recognize that Utah has the youngest population.

☞  To read actual data values, hover over the state. While over Utah you see a median age of 28.5 years. Over Maine, the oldest state, you see a median age of 41.5.

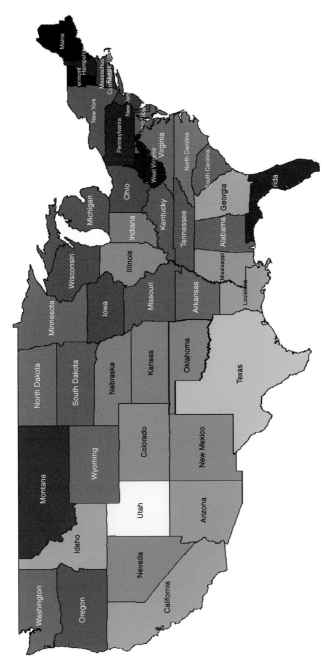

**Figure 2.30** Median Age by State

The color encoding from low value to high value is used to enhance the preattentive effect. It can, however, distort the value assessments. For example, the pale yellow encoding for Utah does not mean that the median age there is close to zero; it just means that it is the lowest value. To avoid such a distortion you can specify that the low value for color encoding is zero.

☞ Check the "Zero Based" radio button in the "Color Encoding" box to make the change.

Now Utah doesn't look quite so young. When viewing the data it is a good idea to do the color encoding both ways. Each has its advantages. Of course, since VisMiner supports multiple displays and view partitions within each display, you could open two boundary viewers for the same data. Use one for "Range Based" encoding and the other for "Zero Based".

☞ Zoom in again by rubber banding the northeast states. Include from Virginia on the south and Ohio on the west. See Figure 2.31.

☞ Check the "Range Based" color encoding and answer the question, "Which northeastern state has the lowest median age?" After a careful comparison you should answer Virginia.

☞ Check the "Relative to Visible" radio button also found inside the color encoding box. Is it now easier to locate the youngest and the oldest states?

**Figure 2.31**   Northeast States – Median Age

Virginia is encoded as a pale yellow because it has the lowest median age of the visible states at 37.1 years.

When you switch back to "Relative to Full Dataset", Virginia suddenly appears older, because the comparison is now relative to all states, which includes Utah.

Again, each encoding has its benefits. As you analyze the presentation be aware of your color encoding selections. Toggle between the options to get a feel for both.

Select the "Zoom Out" button to move back to the previous zoom level.

The "Show Cities" and "Show Highways" check boxes are there for orientation purposes when zoomed in. They do not themselves contain any encoded information.

## Exercise 2.6

Use the USPopDensities.csv dataset to evaluate population distribution by race in the United States. This dataset contains both estimated populations and population densities by state that are broken down by race, ethnic background, and age for 2009.

a. For the 48 contiguous states, which states have the greatest Asian, Black, Latino, NativeAm(erican) and Polynesian populations?

b. For the 48 contiguous states, compare the population counts with the population densities for the Asian, Black, Latino, NativeAm and Polynesian segments. Do this by loading two boundary maps of the same USPopDensities dataset. Use one to show the population for the selected segment. Use the second to show the population density. Discuss your findings.

c. Expand your analysis to include all 50 states. What has changed?

## The boundary data viewer with temporal data

Open the dataset StatePopulationY.csv.

View the summary statistics. As you can see, the dataset contains state populations by year from 1900 through 2009.

Close the summary statistics window.

Drag the dataset to a display.

Select "Boundary Data Viewer".

**Figure 2.32**  Temporal Component Slider

In response to the parameter requests, select "State" as the boundary, and "State" again as the name of the attribute containing state name. Check "Year" as the only temporal attribute.

Select "OK".

After the map loads, zoom to the 48 contiguous states.

You should immediately detect that New York is the largest state (population wise) and Pennsylvania is next. You might question the data if you thought that California had the most people. The answer is found in the new slider in the options panel. The slider is all the way to the left and the label above says: Year 1900. You are looking at the populations in 1900 (Figure 2.32).

To see the populations in subsequent years, slowly drag the slider to the right. As you drag, you see the California color increasing to the more saturated blue color and starting in the 1960s you see New York losing its saturated blue color. This does not indicate that New York is losing population. It indicates that New York's population is dropping relative to the state with the most people. Thus, up until the 1960s, New York is the most populous state and gets the fully saturated blue encoding. Once California passes New York, California gets the fully saturated blue and New York's color encoding becomes relative to California.

Move the slider back to 1900.

Check the radio button "Relative to Full Time Period".

All of the states lighten considerably. This is because the encoding is relative to the maximum over the full 110-year range. The maximum occurred in California in 2009. As with the other color encoding options, you are cautioned to always be aware of the currently selected option. Depending on the dataset, you may want to see the encoding relative to the "full time period" and with a different dataset relative to the currently "selected time".

To animate the progress of population growth over time,

Select the "Start" button.

"Pause" and restart the animation a time or two by repeatedly selecting the button. Do this when you want to take a closer look as the animation progresses.

Exercise 2.7

Use the HousingPriceIndexYQ.csv dataset viewed in a boundary plot to evaluate fluctuations in housing prices. This dataset contains a "Price Index" by state and by quarter, beginning in 1975 and ending in 2010. The index is relative to the state's median home price in 1980 quarter 1. Thus, this data can be used to compare rates of change between states. You cannot use the data to compare actual home prices between states.

a. In 2010, which of the 48 state contiguous states had the largest rate of increase in home prices from 1980? Which had the second largest?

b. In 2010, which of the 48 contiguous states had the smallest rate of increase in home prices from 1980? Which had the second smallest?

c. Using the color encoding only, without hovering to read actual values, determine the year and quarter when home prices in Nevada peaked.

d. Using the color encoding only, without hovering to read actual values, determine the year and quarter when home prices in Hawaii peaked.

e. By 2010, which state had dropped more from peak? Nevada or Hawaii?

f. In 1990 quarter 1, which state's prices were still below their 1980 level?

# Summary

The first two steps in the data mining process are "initial exploration" and "dataset preparation". They are frequently completed in parallel.

VisMiner is a tool that supports both processes using visualizations of the data. The visualizations are designed to preattentively present patterns. Interactions with the visualizations allow the user to eliminate both rows and columns from the dataset, resulting in a more effective and efficient data mining process.

Where advantageous, the visualizations are synchronized. Thus, interactions with one visualization propagate changes in another.

# 3

# Advanced Topics in Initial Exploration and Dataset Preparation Using VisMiner

In Chapter 2, as part of an initial exploration, most of the viewers for data visualization were introduced. At this time, the correlation matrix and the parallel plot were also used to create data subsets. The correlation matrix allowed us to project attributes (dimension reduction) from a dataset, while the parallel plot allowed us to both project attributes and filter observations.

In Chapter 3, although the location plot viewer is introduced, we primarily present additional functionality for dataset preparation. Specifically, we use VisMiner to:

- handle missing values

- create computed columns

- aggregate observations

- merge datasets

- detect and eliminate outliers.

## Missing Values

When working with "real world" data, a common problem is that of missing data. Most analysis algorithms require a complete set of data in order to conduct the analysis. VisMiner is no exception. It requires that all missing values be

*Visual Data Mining: The VisMiner Approach*, First Edition. Russell K. Anderson.
© 2013 John Wiley & Sons, Ltd. Published 2013 by John Wiley & Sons, Ltd.

handled in a dataset before generating visualizations of the dataset or applying data mining algorithms.

Missing values are typically handled in one of five different ways.

- Eliminate any observations with missing data from the dataset. This is usually an acceptable solution when few observations relative to the total number of observations in the dataset contain missing values.

- Keep the observations, but drop the column with missing values. This option may be acceptable when most of the missing values are concentrated in a single column and that particular column is not deemed critical to the planned analysis.

- Assign a default value such as the mean or modal value.

- Use the values of other columns within the dataset to predict a value for the one that is missing. For example, a worker's "age" may be used to estimate the "years of experience" for the worker. The estimated value may not be exact, but at least it allows the rest of the data belonging to the observation to be used in the pending analysis. If the estimate is reasonable, it may not significantly bias the results of an analysis.

- Look in other sources for the missing values before mining the dataset using a tool such as VisMiner. If the missing values are important, it may be worth the effort to find other sources for the missing data that could be merged with your dataset to form a more complete set.

Missing values may occur for two reasons. Either the value is non-existent for the given observation or the actual value is unknown. An example of a non-existent item is the value of the field SpouseName for an unmarried person.

Unknown missing values occur for a number of reasons:

- The entity providing the value was unable or unwilling to report a value.

- The instrument capturing the data malfunctioned during data collection.

- Data may not have been deemed important and thus not collected over a limited span of the data collection period.

Handling missing data in the case of the non-existent value is usually accomplished by assigning a default value. This is especially true when the field type is nominal. In this case, a value of "none" or "not applicable" could be assigned. In the case of numeric fields, assigning a value of zero is appropriate, when it does not conflict in interpretation with data items whose actual value is zero. For example, if in a customer dataset, the field "Total Purchases in Last

90 Days" is missing because the customer has not made any purchases within the time frame, then a value of zero would accurately represent the field.

However, assigning a temperature field the value of zero would bias any analysis results.

## Missing values – an example

The dataset Homes.csv contains data on homes for sale in the Provo, Utah, metropolitan area. The data is from the realtors.com website and contains values entered by realtors when listing a home for sale.

☝ In the VisMiner Control Center, open the Homes.csv dataset. In the icon representing the dataset, notice the red triangle, indicating that the dataset contains missing values and is not ready for analysis or visualization.

☝ View the "Summary Statistics" for this dataset. See Figure 3.1.

Notice at the top of the summary, there are 3,582 rows in the dataset. Also, there are 3,582 rows with missing values. Every single row in the dataset has at least one missing value.

The summary statistics table has a "Missing" column that lists the number of missing items for each of the dataset columns. For example, note that there are

Dataset: Homes.csv

Rows: 3582 (Rows w/ missing values:3582)

| Column Name | Type | Minimum | Maximum | Mean | Std Deviation | Cardinality | Missing |
|---|---|---|---|---|---|---|---|
| bathrooms | Integer | 1 | 11 | 2.79 | 1.189 | 11 | 4 |
| bedrooms | Integer | 1 | 11 | 3.87 | 1.357 | 11 | 16 |
| city | Text | N/A | N/A | N/A | N/A | 51 | 0 |
| cul-de-sac | Text | N/A | N/A | N/A | N/A | 1 | 3191 |
| daysOnMarket | Integer | 1 | 1,468 | 147 | 152.6 | 488 | 0 |
| den | Text | N/A | N/A | N/A | N/A | 1 | 2656 |
| diningRoom | Text | N/A | N/A | N/A | N/A | 1 | 1754 |
| elementary | Text | N/A | N/A | N/A | N/A | 94 | 480 |
| exterior | Text | N/A | N/A | N/A | N/A | 246 | 131 |
| garage | Text | N/A | N/A | N/A | N/A | 11 | 3459 |
| jrHigh | Text | N/A | N/A | N/A | N/A | 21 | 580 |
| latitude | Real | 39.946 | 40.7 | 40.258 | 0.1343 | N/A | 0 |
| laundry | Text | N/A | N/A | N/A | N/A | 1 | 226 |
| longitude | Real | -112.107 | -111.474 | -111.742 | 0.1127 | N/A | 0 |
| lot | Real | 0 | 200 | 0.6 | 5.95 | N/A | 56 |
| mls | Integer | 12,251 | 9,989,816 | 1,066,069 | 635,275 | 3580 | 2 |
| neighborhood | Text | N/A | N/A | N/A | N/A | 1100 | 1209 |
| price | Integer | 14,000 | 8,900,000 | 285,374 | 418,830 | 1002 | 0 |
| propertyType | Text | N/A | N/A | N/A | N/A | 3 | 0 |

Close

**Figure 3.1** Summary Statistics for Homes.csv

four homes listed that do not contain an entry for number of bathrooms. Hopefully, this implies "unknown" rather than "non-existent".

The column "den" has 2,656 missing values while the cardinality is one. Hovering over cardinality for den shows us that the single value is "Y". In this case, we may want to assume that when entering the data, the realtor did not check the "den" box, because there was no den in the home. Therefore we may want to assign "N" those missing values.

There are no missing values for the columns: city, daysOnMarket, latitude, longitude, price, propertyType, SqFeet, state, street, yrBuilt, and zip.

⮕ Close the Summary Statistics window.

⮕ Right-click on the Homes dataset; select "Handle Missing Values".

VisMiner requires that you specify a handling option for all columns with missing data. The "OK" button is disabled (grayed) in this form until you have specified handling options for all listed columns or until you decide to cancel out of the operation.

⮕ Click on the "bathrooms" column. The handling options appear to the right (Figure 3.2). Given that only 0.1% of the observations are missing a

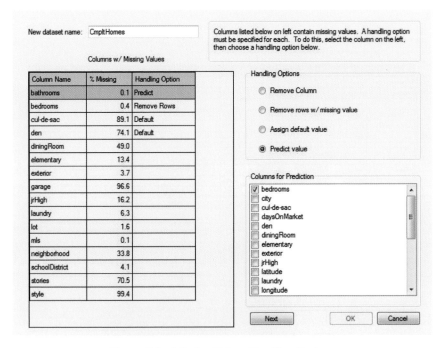

**Figure 3.2**  Missing Values Handling Options

bathroom value, you may want to eliminate those rows with missing values. However, you may also consider using other columns to predict a value. A logical choice of predictors might be the number of bedrooms and the area (square feet) of the home.

☞ Check the "Predict value" radio button. A checked list box appears, allowing you to specify which column values to use as input in predicting missing bathroom values.

☞ Check the "bedrooms" and "sqFeet" boxes.

☞ Click "Next" to advance to the next column.

Predictions of numeric column values in VisMiner are accomplished by building a linear regression model using the checked columns as input variables and the column containing missing values as the output variable. Only those observations containing values for all indicated columns are used to build the model. Once built, the model is then used to generate estimated values for the observations with missing values in the output (predicted) column. To predict missing nominal (text) values, a decision tree is constructed and applied in a similar manner.

☞ Select the "bedrooms" column in the list on the left. For simplicity's sake, select "Remove rows w/ missing value" as the handling option.

☞ Select the "cul-de-sac" column in the list on the left, then select "Assign default value" as the handling option.

☞ In the "Default Value" box enter "N".

☞ Specify handling options for the remaining columns according to Table 3.1.

☞ Click "OK".

Once you have specified the handling option for each of the columns with missing values and pressed "OK", a new dataset named CmpltHomes.csv is automatically created and saved in the same folder or database from which the original was loaded.

☞ Right-click on the original dataset, Homes.csv, then select "Close dataset".

☞ Right-click in the open space in the "Datasets and Models" pane, then select "Reorganize dataset layout".

☞ View the summary statistics for the newly created dataset. Verify that all columns with missing values have been handled.

☞ Close the summary statistics window.

**Table 3.1**  Missing Values Options

| Column | Handling Option | Other |
|---|---|---|
| bathrooms | Predict | use bedrooms and sqFeet |
| bedrooms | Remove rows | |
| cul-de-sac | Assign default | N |
| den | Assign default | N |
| diningRoom | Assign default | N |
| elementary | Assign default | leave blank |
| exterior | Remove column | |
| garage | Remove column | |
| jrHigh | Assign default | leave blank |
| laundry | Assign default | N |
| lot | Remove rows | |
| mls | Remove column | |
| neighborhood | Remove column | |
| schoolDistrict | Predict | use zip |
| stories | Remove column | |
| style | Remove column | |

## Exploration using the location plot

Previously we used the boundary plot to evaluate measures tied to political boundaries when we compared populations by state. In datasets visualized using the boundary plot, there was one row or observation per boundary entity, even though each observation may have contained multiple measure columns.

In this section, we review a related plot – the location plot. Each observation represented in a location plot is a point on a map. To be viewable, the dataset must contain latitude and longitude coordinates. The location plot is similar to a scatter plot layered over a map background. The longitude and latitude values are plotted along the X and Y axes respectively. VisMiner uses a server on the Internet to generate the background map layer. In order to use the location plot, an Internet connection is needed.

☞  If you have not already done so, execute VisSlave on any computers to be used for visualizations.

☞  Drag the newly created CmpltHomes.csv dataset up to a display, release, then select "Location Plot".

☞  A small dialog box opens, asking which column in the dataset contains the observation's latitude and which contains the longitude. In this dataset the latitude column is appropriately named "latitude" and the longitude column "longitude". Select these columns, then click "OK".

The initial location plot view is a map where all observations are plotted as red dots. Map navigation is:

- **pan** – right-button drag

- **zoom** – mouse scroll; as you scroll, the map is automatically centered at the current mouse pointer position.

Notice in the original display, that most of the homes are located around or near Utah Lake. There is one home located to the north and east in the South Snyderville Basin near Park City, Utah. This observation is an outlier. (Outliers are discussed later in the chapter.) It most likely got its location due to a data entry error.

To begin exploration:

☞ move the mouse pointer to a position near Provo Bay on Utah Lake; use your mouse wheel to zoom in a single click. At this zoom level, all observations are visible except for the previously mentioned outlier. You may, however, need to pan slightly to include the upper or lower concentration of points.

As you explore this visualization, you may be tempted to think of the application as a home-finding tool for a potential home buyer. With a few modifications, it could be used for that purpose. Keep in mind that the objective here is data mining. We are looking for patterns in the dataset, not great homes at a bargain price.

In the pane to the right of the map are controls to facilitate the pattern search. (See Figure 3.3.) On top are the category and color encoding options that implement color highlighting of potential relations. Below are the numeric and category filters that allow selective viewing of subsets of data using both the numeric and nominal (category) column values as filters. The numeric filters are double-ended sliders that delineate the range of filtered values. The category filters allow you to include in the display observations having selected values only.

☞ In the Category drop-down, select "schoolDistrict".

☞ To make it easier to read and visually locate, drag the small category key from the upper right corner of the map down over Utah Lake.

Can you readily locate school district boundaries? Although you probably would not use a data mining tool to look up political boundaries, the visualization does provide a quick assessment of the data quality. For example, notice the inconsistencies in school district for homes located up highway 189 in the

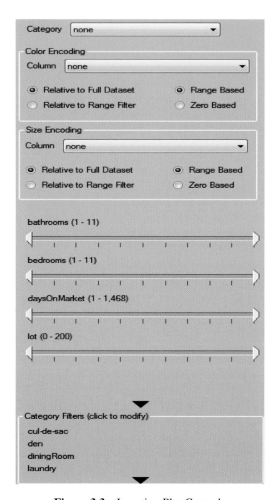

**Figure 3.3**   Location Plot Controls

northeast area of the map. If you are going to use school district as an input variable to a data mining algorithm, you may want to remove these observations from the dataset before performing the analysis. They may bias the results.

In the "Category" drop-down, select "propertyType". Where are the concentrations of "Condo/Townhome . . . " properties?

To see just the "Condo/Townhome . . . " properties, click on the propertyType label within the "Category Filters" box located at the bottom of the pane.

A small dialog opens, allowing you to select which property types you would like to see. Check only "Condo/Townhome . . . " box then click

"OK". Does seeing only "Condo/Townhome . . . " make it easier to locate their concentrations?

☞ In the "Category" drop-down, select "none" to remove the category color encoding.

☞ Click on the "propertyType" category filter, then select "Reset", to clear the filter – once again seeing all observations.

Now look at employing color to encode numeric column values.

☞ In the "Column" drop-down of the "Color Encoding" box select "price".

The dots are color encoded to reflect home price. The darker and more saturated the red, the higher the price. Can you see where the higher priced homes are located?

To assist in locating the higher priced homes, filter out the lower to medium priced homes.

☞ Slowly slide the left end of the price filter to the right. Dragging the range marker to the right filters out those homes whose price is lower than the current position represented by the marker. Also as you drag, the range displayed just above the slider is updated, letting you see exactly what range of prices you are looking at. Drag the left slider up to about the $500,000 price level. You probably won't be able to hit it exactly. What patterns or relationships do you see with respect to price and location?

One problem when color encoding according to price is that a few extremely high priced homes stretch out the full price range. This makes even million dollar homes look relatively pale when encoded.

☞ Eliminate all of the extremely high priced homes by dragging the left price slider all the way to the left.

☞ Drag the right slider down to about the $1,000,000 level.

Even though you are looking at homes ranging in price from $36,000 up to a million dollars, they all look to be very similar in color, because the encoding is still computed relative to the full dataset.

☞ To correct this, click the "Relative to Range Filter" radio button. This color encodes relative only to those homes falling within the current filter ranges. Can you now see a pattern of generally higher prices in the north gradually dropping as you head toward the south? Also notice higher prices on the east, closer to the mountains.

Using the range sliders to filter out higher priced homes does not solve all of the analysis issues with respect to home price. Because the price range slider is scaled to fit the full range of home prices, there is not enough display space to fine tune the slider in the price range up to a million dollars. The needed detailed granularity of the slider is not available. To provide more control when analyzing homes in the under one million dollar price range, create a subset of just those homes.

☞ Ensure that the right end of the price slider is as close to one million as possible.

☞ Select the "Subset" button.

☞ Enter a name of "CheapHomes".

☞ Select "Create".

☞ Close the current location plot.

☞ Open the newly created CheapHomes in the location plot viewer.

☞ Zoom in per previous instructions

☞ Try adjusting the right end of the price slider. Has the precision of the slider improved?

An alternative method for filtering observations from a dataset is available in the ControlCenter.

☞ Right-click on the CmpltHomes.csv dataset.

☞ Select "Create filtered dataset".

☞ Drag the right (high) end of the "price" slider to the left so as to include only homes priced at about $1,000,000 or less (you won't be able to hit it exactly).

☞ Drag the right (high) end of the "sqFeet" slider to the left so as to include only homes smaller than 5,000 square feet.

☞ Under "Category Filters", click on "propertyType".

☞ Check only "Condo . . . ", then click "OK".

☞ For "schoolDistrict", specify "Alpine" and "Provo".

☞ Enter a name of "CheapCondos".

☞ Click "Create".

Although at this time, we will not conduct analyses using the newly created CheapCondos dataset, the filtering option from the Control Center was introduced at this time to allow you to compare this method with other filtering options available in VisMiner using the parallel plot and location plot viewers. For most datasets, the parallel plot is the recommended tool to use when generating filtered subsets. It has the advantage of visual feedback as observations are filtered out. Control Center filtering can be more effective when applied to nominal data types. For example, in the just completed practice example, only homes in the Alpine and Provo school districts were selected. Such a selection is not possible when filtering via the parallel plot where filtering is specified using the sliders, thus requiring that values to be filtered out are limited by adjacency in the plot. Because the nominal values are listed alphabetically, it would be impossible to keep the Alpine and Provo observations while eliminating the Nebo observations.

## Exercise 3.1

Use the CmpltHomes.csv dataset prepared in the previous tutorial.

a. Look for patterns in the relationship between location and year built. What areas have mostly newer homes?

b. When evaluating the relationship between lot size and location, as with price, the few very large lot homes (up to 200 acres) dominate the color encoding. To use the range sliders alone to restrict the selection lacks precision because over 90% of the homes are on less than one acre lots, yet the range slider moves in one acre increments. Thus, in moving the left range "Lot" slider you can't gradually reduce the smaller lot homes. At zero, they are all there, then at the next slider position, they are gone. Use the parallel coordinate plot to first create a subset of the homes having lot sizes less than two acres, then use the location plot to evaluate the relationship between lot size and location.

c. In many areas, proximity to a lake increases a home's value. Does this appear to be the case for the Provo Metropolitan area homes? What geographic setting appears to add value to a home in this area?

## Dataset preparation – creating computed columns

Occasionally, needed or desirable columns in a dataset are not included, but may be computed using other values in the set. For example, suppose that in the CmpltHomes.csv dataset, the relationship between price and location is to be explored. However, looking at price alone is not sufficient, since price is mostly determined by the size of the home. A possible measure representing both price and size is price per square foot – a column that does not exist in the dataset. Let's add it.

⇨ In the Control Center, right-click on the CmpltHomes.csv dataset. Select the option "Create Derived Set".

⇨ Enter a name of "HomeCosts".

⇨ Click "Select All" to include all current columns in the new dataset.

⇨ In the "Computed Columns" box, select "New".

The "New Column Definition" form that opens (Figure 3.4) allows you to define a new column based on values in existing columns. You provide the formula.

⇨ In the "Available Columns" list box, click "price".

⇨ In the "Operators" list box, click the division operator "/".

⇨ In the "Available Columns" list box, click "sqFeet".

⇨ Enter a name of "pricePerSqFt".

⇨ Select "Create", to add the column to the new dataset.

⇨ Select "Create", to build the newly derived dataset.

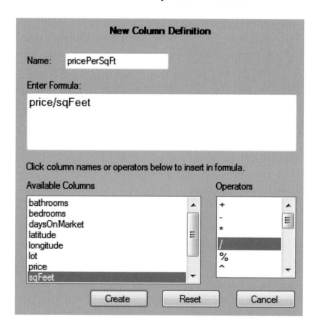

**Figure 3.4**   New Column Definition Form

The formula may be entered directly in the text box, or the list boxes below the formula may be used to interactively build the formula. It is recommended that the list boxes be used, at least to enter column names in order to avoid spelling errors.

☝ Display the new dataset in the location plot viewer.

☝ Slide the left end of the pricePerSqFt range filter up to about 200 in order to remove the typically priced homes.

Where are the remaining homes located? Does there appear to be a relationship between pricePerSqFt and location?

☝ In the Category drop-down, select "propertyType".

Does property type help to explain why sellers of some of these homes would expect a premium price for their home?

## Exercise 3.2

View the newly created HomeCosts dataset in the parallel plot. Hide those columns such as cul-de-sac, elementary, jrHigh, mls, schoolDistrict, street and zip, that are for the most part adding clutter to the visualization. Create two filters – one containing homes with a pricePerSqFt under $200 and the other over $200.

a. Compare the mean values of the two filter sets. What attributes of the data differentiate those homes offered at a high price per square foot?

b. View the dataset in a correlation matrix and a scatter plot. Are there any relationships found using these two viewers that would help to explain the price premium?

c. The correlation between sqFeet and pricePerSqFt slightly positive. In your words, explain why this might be valid, given that in most pricing schemes, unit prices (pricePerSqFt) decrease as the quantity (sqFeet) increases.

## Aggregating data for observation reduction

The dataset ZapataShoes.csv contains records of shoe sales by Zapata Enterprises. Zapata sells three lines of shoes (Fazenda, Montanha, and Praia) directly to consumers in western USA. Each row in the dataset represents a single sale of a pair of shoes. There are almost 500,000 sales recorded in the dataset – too many for the visualization tools. Columns in the dataset include: customer number, customer name, street, city, state, zip, and shoe (line).

☞   Open the dataset ZapataShoes.csv. (Owing to its size, you may need to wait a little longer than normal.)

☞   Review the summary statistics for the dataset.

Zapata would like to explore and compare sales in the geographic areas they serve. In reviewing the available data, there are four columns containing customer location information: street, city, state, and zip. As a starting point, sales could be analyzed by zip, starting with a breakdown by three-digit zip. Since there is no three-digit zip column in the dataset, one needs to be created.

☞   In the Control Center, right-click on the dataset; select "Create derived dataset".

☞   Enter a name of "ShoesZip3".

☞   Click "Select All" to include all existing columns in the new dataset.

☞   In the "Computed Columns" box, click "New".

☞   In the name box of the "New Column Definition" form, enter "Zip3".

☞   In the "Operators" list box, click "trunc" – short for truncate.

☞   Check to ensure that the cursor is between the open and close parentheses.

☞   In the "Available Columns" list box, click Zip.

☞   In the "Operators" list box, click "/" for division.

☞   Type in "100". The formula should now read: **trunc(Zip/100)**. It takes the five-digit zip, divides by 100, then truncates the digits following the decimal.

☞   Click "Create", to add the newly defined column to the dataset.

☞   Click "Create" again, to build the newly derived dataset.

☞   View the summary statistics for your new dataset. There is a new column named Zip3. It has a minimum value of 800, a maximum value of 994, and a cardinality of 154.

There are still almost 500,000 rows in the new dataset – one for each shoe sale. A dataset containing shoe sales totaled by Zip3 is needed.

☞ Right-click on the ShoesZip3 dataset; select "Aggregate rows".

☞ In the "Aggregate On" box of the "Aggregate Dataset Definition" form, check Shoe and Zip3.

☞ In the "Aggregations" box check "Row Count". Each row in the original dataset represents a shoe sale of one pair. "Row Count" will contain the total shoe sale count for each Shoe and Zip3 combination.

☞ Enter a name of "SalesByZip3".

☞ Click "OK".

☞ View the summary statistics for the newly created dataset.

The "SalesByZip3" dataset contains one row for each combination of three-digit zip and shoe – 154 three-digit zips times three shoe lines = 462 total rows.

Viewing the data in the boundary plot would be the logical viewer choice. However, there is still more preparatory work to be completed. The boundary plot requires that the row identifier (key) of the dataset be one of the supported boundary identifiers (state, county, three-digit zip). However, the row identifier of our new dataset is a combination of Zip3 and Shoe.

☞ View SalesByZip3.csv in a parallel plot.

☞ Adjust the "Shoe" filter slider to include only the "Praia" line.

☞ Hide the "Shoe" column. (Right-click on the "Shoe" axis.)

☞ Make a dataset named PraiaSales from the filter. (Right-click on any of the filter sliders.)

The resulting PraiaSales dataset contains two columns: Zip3 and RowCount (sales of the Praia shoe line summarized by Zip3). In other words, almost 500,000 rows, filtered by "Praia" and aggregated by three-digit zip, are reduced down to 154 rows.

☞ View PraiaSales in the boundary plot.

In what geographic areas of the Western States' Zapata market are the Praia shoes selling best?

## Exercise 3.3

Use the previously created SalesByZip3.csv dataset.

a. In the same way that the PraiaSales dataset was created in the previous tutorial, create datasets for Montanha and Fazenda shoe sales.

b.  View each of the datasets in the boundary plot. In what geographic areas of the Western States' Zapata market are these shoe lines selling best?

## Combining datasets

When viewing the PraiaSales dataset in the boundary plot, we see the highest sales volume in the Los Angeles, California (900) three-digit zip. One may ask the question, "Is it highest simply because it has the greatest population?" and "Are there other less-populated regions where the sales rate per capita is higher?"

To answer these questions, data is needed that is not available in the dataset – populations for each of the three-digit zip areas. Population data is readily available from the US Census Bureau and can be downloaded as needed (www. census.gov). The dataset Zip3Pop.csv was previously downloaded. It contains population counts for all three-digit areas in the USA.

☞   Open the file Zip3Pop.csv.

To be useful, the population data in Zip3Pop.csv needs to be combined (or joined in database terminology) with the sales data in PraiaSales. The VisMiner Control Center has a feature that implements dataset joins. The join feature requires that each dataset contains one or more key columns that match a row or rows in the other set. In the case of the PraiaSales and Zip3Pop.csv datasets, each has the common Zip3.

☞   Drag the PraiaSales dataset over the Zip3Pop.csv and release.

☞   Select "Join datasets".

☞   A dialog opens requesting specification of the column names in each dataset to be used to match rows in the other dataset. Check the Zip3 column listed under the Zip3Pop.csv header on the left.

☞   Check the Zip3 column listed under the PraiaSales header on the right.

The Zip3 column in Zip3Pop.csv is to be matched with the Zip3 column in PraiaSales. In this case, the two columns have the same name. This is not a requirement. They just need to originate from the same source domain. The join may also require multiple key columns to perform the join. For example, if instead of three-digit zips, county populations were to be combined with interesting county data collected in a different dataset. Since county names in the USA are not unique, one would need both a county name column and a state name column in each dataset in order to successfully match rows.

☞   Click "OK" to join the two datasets.

A base dataset is created with the combined name of "Zip3Pop_join_PraiaSales".

↪ View the summary statistics for Zip3Pop_join_PraiaSales.

The new dataset contains three columns: Population from the Zip3Pop dataset, RowCount (total sales) from the PraiaSales dataset, and Zip3 which was common to both. Before viewing the dataset, an additional column is needed: SalesPerCapita.

↪ Right-click on Zip3Pop_join_PraiaSales.csv; select "Create derived dataset".

↪ Name the new dataset "PraiaPerCapita".

↪ Click "Select All" to include all existing rows.

↪ Click "New" to define a new computed column.

↪ Name the new column "SalesPerCapita".

↪ Enter a formula of "RowCount/Population".

↪ Click "Create" to create the new column.

↪ Click "Create" to create the new derived dataset.

↪ View the new dataset "PraiaPerCapita" using a Boundary Plot.

↪ In the "Show data for" drop-down, select "RowCount (total sales)"; review the results.

↪ Switch to SalesPerCapita. Again, review the results.

In SalesPerCapita, we see a more accurate picture of areas where the Praia shoe line is selling well. High sales rates exist in all of coastal southern California as well as southern Nevada, southern Arizona, and southern New Mexico. We even see a high sales rate in Oregon around the Eugene area. This could not have been done without first joining the private Zapata sales data with the publicly available census data.

## Exercise 3.4

In the same way that a new dataset was created to evaluate Praia sales per capita, evaluate sales for the Fazenda and Montanha lines.

a. Where are Montanha shoes selling best?

b. Where are Fazenda shoes selling best?

c. In which regions do multiple lines sell best?

## Outliers and data validation

An outlier is an observation containing column values or a combination of values lying outside the expected range of the population from which the dataset observations were drawn. When outliers are included in datasets applied to data mining algorithms, they may bias or invalidate the results of the analysis. Hence, before application of the algorithms, the outliers should be identified and either corrected or removed.

Outliers are usually generated in the data collection process. The measuring instrument may have been faulty; there may have been errors in data entry; or there may have been errors in communications or transmission.

Occasionally, the outlying entries may be valid, yet drawn from a different population than the rest. These values may be explained by additional attributes not captured in the dataset. For example, employers in a given region frequently share their salary information in order to help each other understand how their salaries compare to those of other employers in the region. In a study of "custodian" salaries, most employers reported salaries in the range of $20,000 to $40,000 per year. One employer, however, reported a salary of $95,000. When contacted to verify the salary, the employer responded, "Yes, that is correct, this is my father-in-law, a part-owner of the company. He was given the title of 'custodian', because he has assumed the responsibility for taking out the trash and vacuuming the floors." In this case, the data was correct, but the definition of "custodian" was not the same as that accepted by other submitting employers.

In the search for outliers there are a number of checks that may be applied.

- **Range** checks – values that are outside the expected range for a given attribute.

- **Computed** checks – if there is redundancy in the dataset, is there consistency in the redundancy? For example, does the total of all probabilities sum to one?

- **Feasibility** and **consistency** checks – are all attribute combinations possible? For example, is it possible to have a female patient diagnosed with prostate cancer?

- **Pattern** checks – when there are patterns observed between attributes in a dataset, do all observations reasonably fit those patterns?

- **Temporal** checks – in datasets containing multiple observations of the same entity over time, are there large unexpected changes in a given attribute value for a single entity?

The search for outliers can be a time-consuming operation – but it is important. In the examples that follow, visual approaches are presented using VisMiner. They enable the search for outliers and support a more efficient process.

Working with outliers is typically a two-step process.

1. Locate and identify.
2. Handle by
    a. removing from the dataset; or
    b. correcting the data.

In the examples that follow, VisMiner is used to visually locate outliers. Once they are identified, the parallel plot is used to isolate the outliers and eliminate when appropriate. VisMiner does not have a mechanism for correcting data. This step must be completed outside of VisMiner.

## Range checks

In doing range checks, we are searching for values outside the expected range for a given column.

The range may be predefined by a set of limits (a test score must be between 0 and 100 inclusive), or it may be less rigidly defined as the range of expected values for an item with a known or assumed distribution.

## Fixed range outliers

When attributes in datasets have known fixed ranges, location of invalid values is quite simple. The dataset ProspectiveStudents.csv contains data on students applying for college admission. It includes:

- HighSchoolGPA – in this dataset, it should be between 0 and 4.0

- CompACT – the prospect's composite ACT score; zero, if the test has not been taken. Valid scores range from 1 to 36 inclusive.

- SAT_CRM – the prospect's combined comprehensive reading and math SAT score; zero, if the test has not been taken. Valid scores range from 2 to 1600 inclusive.

Let's use this range information to search for outliers.

⑤ Open ProspectiveStudents.csv.

⑤ View the summary statistics.

The CompACT column has a maximum score of 60. This is a problem. There are also invalid values in the HighSchoolGPA and SAT_CRM columns as evidenced by their maximums. The invalid values should be either removed or corrected. To remove:

☞   View ProspectiveStudents.csv in a parallel plot.

☞   Drag the top filter sliders down to eliminate the visible offending observations on the HighSchoolGPA, CompACT, and SAT_CRM axes.

☞   Right-click a range slider; select "Make dataset from filter".

☞   Name the new dataset "ValidStudents".

☞   Click "Create".

You now have a dataset containing only valid observations. If instead of removing the invalid observations, you would prefer to correct the problems, you will first need to identify observations containing invalid data.

☞   Drag "ValidStudents" up over "ProspectiveStudents.csv" and drop.

☞   Select "Create dataset from difference".

The difference subset "ProspectiveStudents.csv-ValidStudents" includes all observations with invalid data. To correct, outside of VisMiner, use the IDs in this set to locate the observations in "ProspectiveStudents.csv". Go back to the original sources to find correct values for these prospects and reenter.

## Distribution based outliers

If a normal distribution of values is assumed, we may arbitrarily classify all values beyond a predetermined number of standard deviations from the mean as outliers. Yet using this rule, depending on the dataset size, there may be expected valid observations whose values fall outside the defined permissible range.

For example, assuming a normal distribution, the probability that an observation will fall more than three standard deviations from the mean is 0.0027. Thus, in a sample of one thousand observations, two or three observations would be expected outside the three standard deviation range. If the assumption with respect to the distribution is correct, these would, in fact, be valid observations. Even if an outlier is defined as anything beyond five standard deviations from the mean, in a dataset of 100,000 observations, six or seven valid observations outside the acceptable range would be expected.

Rather than set range limits, then deterministically identify all observations outside the acceptable range, a more efficient method is to visually inspect each of the attribute distributions. The human visual system is very good at preattentively locating outliers. In the search for outliers using VisMiner, begin with the parallel plot, because of its ability to concurrently plot more than three dimensions and to quickly eliminate outliers using the range sliders.

⮑  View the dataset Table1.csv in a parallel plot.

⮑  By rotating the display and dragging each axis out to the left or right ends of the plot, evaluate the distributions of each column.

The columns named Y and Z look to be normally distributed. The ID column, as expected by its name, has a very smooth uniform distribution, while the U column appears uniform combined with some random variations. At the low end of the Y axis there is an outlier. In evaluating the Z axis, one may question the high value of 20.3 due to the separation between it and the other Z values. Based solely on a visual judgment, however, the separation is probably not enough to be classified as an outlier.

⮑  Eliminate the outlying observation by dragging up the bottom slider on the Y axis until it disappears.

⮑  Right-click on the slider; select "Make dataset from filter".

⮑  Give the new dataset a name of "Table1Valid".

At this point in the process, the table has been visually scanned, an outlier has been found, and a new dataset has been created without the offending outlier. Although nothing more will be done with the dataset in these tutorials, it may be considered ready for mining.

⮑  Open the dataset Table4.

⮑  Examine the distribution of var1 using the parallel plot.

As with the Y column of Table1, the var1 distribution looks normal except for the 19 observations falling below the bell-shaped distribution above. If there were just one or two observations in this range, then they could be classified as outliers. However, with 19, they are very unlikely to be just data collection errors. There must be an explanation for their existence.

⮑  Drag the top var1 filter slider down to eliminate all but the bottom 19 observations.

⮑  Add another filter to the plot.

⮑  Click on the lasso for the second (blue) filter. Rubber band the lasso over the var1 column to select all but the bottom 19 observations.

⮑  Check "Show Means".

The values of var2, var3, and var4 look to be similar between the filter group containing the bulk of the data and the one containing the 19. With var5, there is a greater difference. Although in this tutorial the reason for that difference will not be pursued, when conducting a complete analysis, further investigation is needed to assess the relationship between var5 values and the 19 that fall outside the expected distribution with respect to var1.

For an example of sub-population-based outliers, let's return to the iris dataset.

⌖   View in a parallel plot the iris.csv dataset introduced in Chapter 2.

With respect to PetalLength and PetalWidth, a sub-population of observations is visible. This sub-population was previously found to be explained by Variety – they are all Setosa.

⌖   Create three filter groups – one for each variety.

⌖   One at a time, examine the three groups by hiding the other two.

When looking at just the Setosa variety, a potential outlier is visible near the bottom of the SepalWidth axis. It should be further investigated.

This last example illustrates the need to separately evaluate sub-populations within a dataset when those sub-populations are known *a priori*. In VisMiner, the task is facilitated using the filter capabilities of the parallel plot.

## Computed checks

When columns in a dataset may be derived from other columns, the validity of all involved columns may be confirmed, although the identity of the offending column may not be obvious.

Player statistics released by Major League Baseball provide an example. The dataset mlbBatters2011.csv contains the end of season batting statistics for the 2011 season. The dataset as released was valid. However, in the dataset employed here one observation has been altered in order to illustrate the process.

⌖   Open the dataset mlbBatters2011.csv and view the Summary Statistics.

To a domain expert familiar with baseball statistics, the following derivations are identified:

$$BattingAvg = Hits/AtBat$$

$$TotalBases = Hits + 3 * HomeRuns + 2 * Triples + Doubles$$

$$Slugging = TotalBases/AtBat$$

To confirm the validity of column values involved in the computation of TotalBases:

⇪ Right-click on mlbBatters2011.csv; select "Create derived dataset".

⇪ Name the new dataset mlbCheck.

⇪ Click "Select All".

⇪ Define a new column named "TotalBasesChk" with a formula of:

$$\text{Hits} + 3 * \text{HomeRuns} + 2 * \text{Triples} + \text{Doubles}$$

The above new column uses the four inputs (Hits, HomeRuns, Triples, and Doubles) to recompute TotalBases. In order to not conflict with the existing TotalBases column, it is given the name "TotalBasesChk".

⇪ Define a new column named "TotalBasesDiff" with a formula of:

$$\text{TotalBases} - (\text{Hits} + 3 * \text{HomeRuns} + 2 * \text{Triples} + \text{Doubles})$$

If all entries are valid, the values of the existing TotalBases column and those of the computed TotalBasesChk should be identical. Thus the difference between the two (TotalBasesDiff) should be zero.

⇪ Click "Create" to create the derived set.

⇪ View mlbCheck in a parallel plot.

Look at the connecting lines between TotalBases and TotalBasesChk. When all are horizontal, it is a good indication that the recorded values are correct. A stronger statement of correctness can be made when all connecting line segments are horizontal, the minimum values match, and the maximum values match.

If all the data is valid, the difference between the recorded value and the computed value should be zero. In the plot, focus attention on the column TotalBasesDiff – the difference between the recorded value (TotalBases) and the computed value (TotalBasesDiff). To reduce clutter, all but the relevant columns (Doubles, Hits, HomeRuns, Triples, TotalBases, and TotalBasesChk) may be hidden.

In looking at the column TotalBasesDiff. All but one of the differences are zero. In the parallel plot, that loner stands out as an outlier. The difference cannot be explained by rounding, since the computation includes integer addition and multiplication only. There is definitely an error in at least one of the involved column values.

⊖  If columns have previously been hidden to reduce clutter, make them visible again. A subset is going to be created of the outlying observation. All columns need to be included in that subset.

⊖  Drag the bottom slider of the TotalBasesDiff column up until none but the outlying observation is visible.

⊖  Right-click on the slider; make a subset from the filter named "ComputedOutlier".

⊖  View the data for ComputedOutlier in a table (tabular presentation).

The player's name is Granderson, an outfielder for the Yankees. In comparing the data with the original source, it was found that Granderson actually hit 41 home runs rather than 14. The digits were transposed at data entry. To proceed, one can choose to either find and enter correct values or remove the observation from the dataset. If the choice is made to remove, do so by subtracting the outlier dataset from the base dataset (mlbBatters2011.csv). Correcting values in a dataset is outside the scope of VisMiner. It can easily be done using any of a number of "csv" file editors, such as Microsoft Excel or Windows Notepad.

## Exercise 3.5

Using the dataset mlbBatters2011.csv, create a subset adding computed columns for BattingAvgChk, BattingAvgDiff, SluggingChk and SluggingDiff. Take the same approach as was done in the tutorial for the TotalBases data. In baseball statistics,

$$\text{batting average} = \text{hits}/\text{at bats, and}$$
$$\text{slugging} = \text{total bases}/\text{at bats.}$$

If computed correctly, the values for BattingAvgDiff and SluggingDiff are non-zero. Does this indicate that there are errors in the data? Explain your answer. (Hint: Look at the magnitude of the differences.)

## Feasibility and consistency checks

The dataset Amarillo.csv contains data on homes for sale in Amarillo, Texas. It was generated from on-line homes-for-sale listings by Amarillo realtors. The data includes the latitude, longitude, and county name as entered by the realtors. An interesting characteristic of Amarillo is that it is split in half by two counties:

Potter and Randall. The border between the two runs along latitude 35.18. To the north is Potter County and to the south is Randall County.

↪ View in a parallel plot the Amarillo.csv dataset.

↪ To allow you to focus on the relationship between County and Latitude, hide all but these two axes.

↪ Drag the top Latitude range slider down until it reaches 35.17. Since there is some leeway in slider placement at the 35.17 level, try to keep it as high as possible while still at 35.17.

Homes below 35.17 degrees latitude are definitely in Randall County even though four that are close to the dividing line are listed in Potter County.

↪ Drag the top Latitude range slider back to the top, then drag the bottom range slider up until it reaches 35.19.

Homes above this latitude are in Potter County even though 41 are listed in Randall County.

There is definitely incorrect data in this dataset. Either the county is listed wrongly or the location coordinates are invalid. If location or county are important in the planned data mining analysis, the errors must be isolated and corrected or removed from the dataset.

Note: Given the discrepancy in number between the Randall County homes assigned to Potter County (4) and the Potter County homes assigned to Randall County (41), we make the assumption that the county values are in error. This is based on domain knowledge that Randall County is generally a more desirable county for residential real estate. Realtors making the entries likely coded the county as "Randall" in order to have the homes show up in searches for Randall County homes.

## Data correction outside of VisMiner

In a previous example, presenting the handling of missing values and intro- ducing the location plot viewer, we explored the homes-for-sale data in the Provo, Utah, metropolitan area (Homes.csv). That data had been downloaded from the realtors.com website. The dataset as downloaded contained numerous missing values, which were handled in the tutorial. Although not discussed in the tutorial above, the downloaded data also contained questionable locations. These had previously been corrected before its use in the tutorial.

Let's now go back and look at that original data as found in UtHomesAs- Downloaded.csv. (Note: In order to focus on the location issues, the missing values have already been handled. For the most part, the columns with missing values were eliminated.)

👉   View UtHomesAsDownloaded.csv in a scatter plot.

👉   Select Longitude for the X axis and Latitude for the Y axis.

👉   Rotate as needed to view the height densities.

   A scatter plot with longitude and latitude on the axes is essentially a location plot without the background map layer. A scatter plot was selected because of its ability to plot density histograms (Type = Height Density). In reviewing the plot, notice the highest concentration of homes for sale around latitude 40.406 and longitude −111.883. Given domain knowledge of the metropolitan area, the validity of homes concentrated in that small area was questioned. Using the parallel plot and the text-based table view, we zoomed in on those homes to find 123 entries with the exact same latitude and longitude; yet the street addresses were all different. Apparently, the realtor entering the listing did not have the latitude and longitude of the property and did not bother looking it up. Instead they entered a generic location. In checking other areas of high concentration, the same problem was found although not to the same extent.

   To correct the problem outside of VisMiner, publicly available geocoding services were used. Geocoding is the process of converting a street-city-state-zip address into latitude and longitude coordinates. Map serving sites such as Google, Yahoo, and Microsoft include free geocoding services in limited quantities. The University of Southern California GIS Research Laboratory (webgis.usc.edu), which was used for these entries, provides free geocoding for up to 2,500 addresses. It applies usage-based fees beyond that level.

## Distribution consistency

The dataset BodyTemp.csv contains readings of patient body temperatures as collected by nurses in a hospital over a two-week period.

👉   In the dataset BodyTemp.csv, view the temperature column distribution in a histogram.

   The temperatures are recorded as integers in tenths of a degree Fahrenheit. In the distribution, we see the opposite of what we saw in the UtHomes dataset. Instead of an unusually high concentration of observations, we see a lower than expected number of normal (98.6 F) patient temperatures. Upon investigation, the hospital found that nurses were reluctant to report a temperature of 98.6 as it might lead a supervisor to question whether the temperature had actually been taken. When a reading of 98.6 was taken, nurses in recording the temperature, frequently bumped it up or down a tenth of a degree to give the appearance that it had actually been taken, yet remained well within the "normal" range.

If this data is to be used in a data mining application, depending on the methodology applied, it may be prudent to reassign some of the 98.5 and 98.7 temperatures back to 98.6. Algorithms such as artificial neural networks will pick up on this distribution anomaly and distort the descriptive abilities of the resulting models.

## Pattern checks

With respect to patterns, the search is for outlying observations that do not conform to the typical relationship between two or more attributes. The correlation matrix coupled with the scatter plot are an effective combination of viewers in the visual search for these outliers. Begin with an exploration of a synthetic dataset.

⮂  View Table6.csv in a correlation matrix.

The correlation matrix is a good starting point in the identification of attribute relationships. Using the correlation matrix we locate pairs of correlated attributes, then in a synchronized scatter plot we evaluate the relationship. In Table6 we see just one candidate – var1 versus var2.

⮂  View Table6.csv in a Scatter Plot.

⮂  In the correlation matrix, click on the var1-var2 cell to bring it up in the scatter plot. (See Figure 3.5.)

Do you see the outlier? Notice that if one were to look independently at distributions on var1 and var2, the outlier would not be found. It is hidden within the single-dimension distributions of each. The outlier only surfaces when the two-dimensional plot pairing both attributes is examined.

As a matter of practice, in conducting pattern checks, open the dataset in question in both a correlation matrix and a synchronized scatter plot. Systematically click from cell to cell in the correlation matrix on attributes combinations with potentially meaningful correlations, examining the corresponding scatter plot as you progress.

Visually locating outliers based on multi-attribute relationships, is also possible in the parallel plot.

⮂  View Table6.csv in a parallel plot.

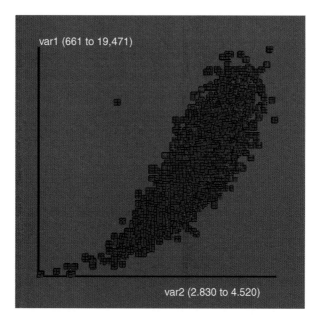

**Figure 3.5**   Scatter Plot for Table6.csv

A cursory review of the plot does not reveal any obvious outliers.

☞   Rotate the plot to reveal its back side by dragging horizontally. (See Figure 3.6.)

Do you see the outlier? Although not as preattentive as in the scatter plot, the outlier is visible. It is represented by the connecting line between var1 and var2 that runs in the opposite direction from the other observations. (Figure 3.6.) If you have trouble locating the outlier, it may help to watch as you rotate the plot slightly.

Note that in the parallel plot, pattern-based outliers are only visible between adjacent columns. To view all possible combinations, systematically reorder the columns. A systematic search is more effective using the correlation matrix/-scatter plot combination rather than the parallel plot. We present the parallel plot view of outliers for two reasons. First, as you are using the parallel plot for other exploration purposes, you may run across pattern-based outliers and should be able to recognize them. Second, the parallel plot is effective in isolating and removing the outlier.

☞   Drag the top var2 range slider down, approaching the outlying observation. Getting really close is not necessary; just get close enough to see the outlier clearly isolated on the var1 axis.

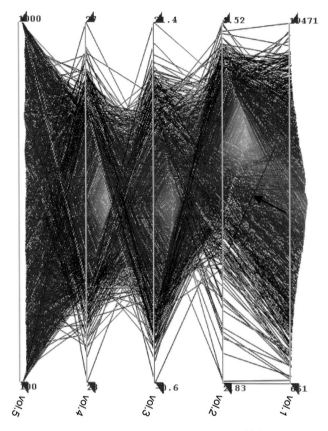

**Figure 3.6** PCP of Table6.cvs Back Side

☞ Drag the bottom var1 range slider up until only the single outlying observation is visible.

☞ Right-click on a range slider, selecting "Make dataset from filter" to create a subset containing only the outlying observation.

☞ Name the dataset "Table6Outlier".

At times when removing outliers using the parallel plot it is easier to create a subset of all but the outliers; and at other times, in Table6 for example, it is easier to create the subset of just the outlying observations. When this is the case another step is required using the Control Center's "Difference" operation.

☞ In the Control Center, drag Table6Outlier over Table6.csv and drop.

☞ Select "Create dataset from Difference".

After subtracting (or removing) the outliers, the resulting dataset contains just valid observations, which was the objective. The dataset name assigned by the Difference operation is a combination of both involved datasets. In this case it is "Table6.csv-Table6Outlier". A better shorter name might be preferred.

☞   Right-click on Table6.csv-Table6Outlier; select "View/Edit name and notes".

☞   Change the name to "Table6Valid".

☞   Click "Save".

## A pattern check of experimental data

The dataset ResponseTime.csv contains the results of benchmark tests comparing to widely used web servers, identified in the dataset as "Platform A" and "Platform B". The data was collected by using a simulator to repeatedly make requests of web pages from the servers from hundreds of different locations then measuring the response times.

☞   Open ResponseTime.csv.

☞   View the Summary Statistics.

The AvgPgRsp is the average time in milliseconds that it took the server to respond, given the requested page size (FileSize in kilobytes) and the number of requests per second (TPS) hitting the server. The average was based on thousands of requests to the server from hundreds of clients requesting files of the specified size and the given traffic level. As you see from the summary statistics, file size requests ranged from 5 KB up to 50 KB and traffic levels ranged from 100 to 550 requests per second.

☞   View ResponseTime.csv in a scatter plot.

☞   Select TPS on the X axis and FileSize on the Y axis.

In this view, we clearly see the benchmark design. This is not randomly sampled data, but a carefully crafted experiment. The apparently missing observations in the upper right corner of the grid (A in Figure 3.7) represent trials that overloaded the server so much that it failed to respond to the page requests. The missing observations in the column at 225 TPS (B in Figure 3.7) were unintentional omissions made by the lab technician running the

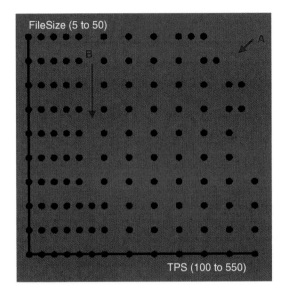

**Figure 3.7** Scatter Plot of ResponseTime.csv

simulations. These omissions were not detected until results were viewed in a similar scatter plot.

Change the Y axis to AvgPgRsp, the Z axis to FileSize, and the category to Server.

Numerous outliers are visible among the observations at the 250 TPS level. After review, these turned out to be simulations run without first returning the servers to a predetermined initial state – again a mistake made by the technician. They obviously need to be removed from the dataset before continuing the analysis. The higher page response rates at the upper TPS and FileSize settings were valid measurements reflecting degradation of the server at these levels. The phenomenon observed in the plot is frequently referred to as the "hockey stick".

## Exercise 3.6

Using the ResponseTime.csv dataset in VisMiner:

a. Use the parallel plot to extract a subset named "outliers" of the invalid observations.

b. In the Control Center, create a subset of valid observations using the difference between the full dataset and the outlier set.

c. Name the dataset "ValidResponseTime.csv".

# Summary

In preparation for the application of data mining algorithms, most datasets need some modification. These modifications include:

- projection – attribute selection
- restriction – filtering of observations
- sub-population extraction
- aggregation – combining of observations
- elimination of missing values
- derivation of new columns
- merging of datasets
- detection and elimination of outliers.

VisMiner supports all of these operations. Where suitable, some are supported within the parallel plot, correlation matrix, and location plot viewers. Other modification operations are implemented directly in the Control Center.

# 4

# Prediction Algorithms for Data Mining

In support of the data mining process, VisMiner implements algorithms for prediction modeling. It supports modelers both for classification (predicting nominal or class values) and regression (predicting continuous numeric values). In this chapter we introduce the basic algorithms implemented by VisMiner. These include decision trees, support vector machines, and artificial neural networks for classification and artificial neural networks for regression.

For the most part, the algorithms of VisMiner are a black box. One does not need to know precisely how the algorithms work in order to deploy them in data mining exercises. Consequently, this chapter may be skipped. However, knowledge of the algorithms can help in the following ways:

- Algorithm selection – each algorithm has its strengths and weaknesses. An understanding of the internal workings of an algorithm leads to a better appreciation of its strengths and weaknesses. Consequently it results in better decision making when it comes to algorithm selection as dictated by the dataset characteristics and data mining objective.

- Results evaluation – knowing how the algorithm arrived at its results helps in assessment of the applicability and confidence in the results. For example, with respect to a decision tree, how does a root level split variable compare in importance to a leaf level split?

*Visual Data Mining: The VisMiner Approach*, First Edition. Russell K. Anderson.
© 2013 John Wiley & Sons, Ltd. Published 2013 by John Wiley & Sons, Ltd.

There are issues in the application of all algorithms implemented by VisMiner. When choosing an algorithm to apply to a dataset, ask the following:

- What modeling parameter specifications are required to use the algorithm? Are there parameter specification guidelines?
- Are there limitations on the dataset size and data types?

  o Does it scale well?

  o If nominal types are supported, is there a cardinality limitation?

  o If nominal types are encoded as numeric values, will the resulting models be distorted?

- Does the algorithm guarantee that it will arrive at the same result for a given dataset or is there a random component to the process that may not always lead to convergence?
- Does the algorithm guarantee an optimal solution or, during solution search, might it stop at local optima?
- Is the resulting model easy to interpret and understand?
- Does the algorithm support incorporation of analyst domain knowledge before or during the process?
- What are the measures of model performance and possible tests for model significance?
- How likely is the algorithm to overfit a model to the data?

## Decision Trees

**Decision trees** are relatively simple computing methodologies for classification, that generate a tree structure of training data subsets by recursively splitting a dataset on selected input attributes. The algorithm is as follows:

1. Begin with a single node composed of all observations in the training set.
2. Choose the best input attribute whose values within the node can be used to split the node into subsets that are more homogeneous with respect to the classification attribute than the unsplit node.
3. For each subset of the split, determine if that subset can or should be split. For each subset to be split, return to step 2.

The above description leaves three questions unanswered:

1. How is "more homogenous" defined?
2. How should the best input attribute be chosen for a split?
3. When should a node not be split?

A simple index or measure of homogeneity is classification error, defined as one minus the fraction of the most frequently occurring class. For example, if in a binary classification node, class A occurs 40% of the time and class B occurs 60%, then the classification error is 0.40. If there are three class values (A, B, and C) occurring with frequencies of 45%, 30%, and 25% respectively, then the classification error is 0.55.

In a binary classification node the classification error is in the (0, 0.5) range, where error is zero if all observations within the set belong to the same class. In a ternary classification node the range is (0, 0.667).

An alternative index of homogeneity is the Gini index or coefficient. It is computed as:

$$Gini = 1 - \sum_{i=1}^{c} p(i)^2$$

where

$c$ is the number of unique class values
$p(i)$ is the fraction of class $i$ in the node.

Like the classification error, the range in a binary classification is (0, 0.5).
To measure the benefit of a proposed split, we compute the difference between the index of the parent node and the weighted average of the child node indices. The formula for gain is:

$$Gain = I_{parent} - \sum_{j=1}^{k} \frac{N_j}{N} I_j$$

where

$I_{parent}$ is the computed index of the parent node
$I_j$ is the computed index of child node $j$
$k$ is the total number of child nodes
$N$ is the number of observations in the parent node
$N_j$ is the number of observations in child node $j$.

With a computation for gain, we are now ready to answer the second question above – "How should the best input attribute be chosen for a split?" The simple answer is to choose the input attribute and split that maximizes the gain.

The type of split depends on the data type and distribution of the input attribute. For example, suppose that input attribute W is nominal with two different values (X and Y). The only possible split is a binary split placing all X observations in one child node and all Y observations in the other. However, if W has three different values (X, Y, and Z), the possible splits are:

- a binary split with all X observations in one node and all Y and Z observations in a second node

- a binary split with all Y observations in one node and all X and Z observations in a second node

- a binary split with all Z observations in one node and all X and Y observations in a second node

- a ternary split with all X observations in one node, all Y in a second node, and all Z in a third.

The number of alternatives exponentially increases as the cardinality of the nominal input attribute increases. To avoid this complexity, an acceptable rule of thumb is to only consider binary splits.

When the input attribute is continuous (numeric) and only binary splits are considered, a split position must be determined such that all observations whose input attribute value is less than or equal to the split position are assigned to the first child node and all other observations assigned to a second node. Hence, the problem becomes locating the split position that maximizes the gain for the input attribute under question. Depending on the number of unique values for the input attribute in the node, this could become a rather exhausting search. To simplify, one alternative is to choose a finite number of equally spaced split positions (10 for example), evaluate each – choosing the position yielding the greatest gain.

The algorithm to choose the best split becomes:

1. For each potential input attribute, find the split based on that attribute yielding the best gain.

2. Choose the input attribute with the overall best gain.

### Stopping the splitting process

This brings us to the final question of when to terminate the splitting process. There are a number of possible "stop rules" that may be applied:

1. A node containing only one class value (homogeneity measure of zero) should not be split.

2. A node containing identical input values on all input attributes cannot be split.

3. A node whose best possible split gain is below a threshold is not worth splitting.

4. A node split resulting in any child nodes whose size is below a minimum size threshold should not be split as it is likely to produce a model that does not generalize.

5. A node split with an acceptable gain with respect to the training dataset but with negative gain with respect to the validation dataset will produce an overfit model.

When training, all of the above should be considered. The first two are mandatory. It is good practice to use a validation dataset to help decide when to stop splitting.

### A decision tree example

Consider the dataset in Table 4.1. A car dealer has collected data on 10 visitors to its showroom. The classification problem is the construction of a decision tree that uses Gender, MaritalStatus, and SpeedingCitations columns to predict CarBuyer. In this example, a decision tree is manually built by following the previously described algorithm.

The first step is to determine the best attribute on which to split. In calculating gain, we use the classification error as the homogeneity index. We use this rather than the Gini index because it is simpler to compute and will allow you to follow

**Table 4.1**   Car Buyer Data

| CarBuyer | Gender | MaritalStatus | SpeedingCitations |
|----------|--------|---------------|-------------------|
| Y | M | S | 5 |
| Y | F | S | 3 |
| N | F | M | 0 |
| N | M | M | 0 |
| Y | M | S | 3 |
| N | F | M | 1 |
| Y | F | M | 1 |
| N | M | S | 3 |
| N | F | M | 4 |
| Y | M | M | 3 |

along doing calculations in your head. The classification error of the full dataset is 0.5 since there are an equal number of buyers (Y) and non-buyers (N).

- A split on Gender would create a node of male (M) visitors with 3 buyers and 2 non-buyers. The other node would be female (F) visitors with 2 buyers and 3 non-buyers. The index of each node is 0.4, resulting in a gain for the split of 0.1 (0.5 − 0.4).

- A split on MaritalStatus would create a node of single (S) visitors with 3 buyers and 1 non-buyer. The other node would be married (M) visitors with 2 buyers and 4 non-buyers. The index of the first is 0.25 and of the second is 0.33, resulting in a gain for the split of 0.2 (0.5 − 0.3).

- A split on SpeedingCitations at 2 (the best available) would create a node of speeders (> 2) with 4 buyers and 2 non-buyers. The other node would be slow folks (<2) with 1 buyer and 3 non-buyers. The index of the first is 0.33 and of the second is 0.25, resulting in a gain of 0.2 (0.5 − 0.3).

Since the gain of the MaritalStatus and SpeedingCitations splits are equal, either could be chosen. Figure 4.1 shows the decision tree after splitting on SpeedingCitations.

Focusing on the left node, an evaluation of gain weakly recommends splitting on either Gender or again on SpeedingCitations. We say weakly, because using classification error as the index, the gain is 0.0 for each. When using the more complex Gini index, both potential splits generate a positive gain. The gain of SpeedingCitations is greatest. Although not detailed in the example, the same splitting methodology could also be executed to process the right node at the second level. The resulting tree is in Figure 4.2 after a split using Speeding-Citations on the left and MaritalStatus on the right.

In Figure 4.2, the bottom node on the left must be terminated because it only contains non-buyers − stop rule 1. The second node from the left must also be terminated because the input attributes are identical − stop rule 2.

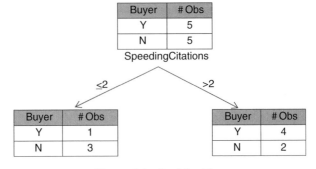

**Figure 4.1**  Decision Tree

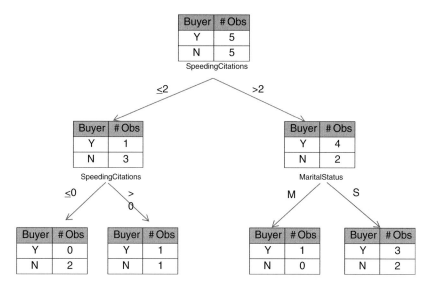

**Figure 4.2**   Decision Tree after Splits

## Using decision trees

To make a prediction using input data without a corresponding output class, find the leaf (terminating) node in the tree matching the input values. For example, suppose that in the example above, we have a single, male showroom visitor with three speeding citations. The leaf node for this person is the lower right node of Figure 4.2. The predicted class value is the most frequently occurring class in the node. The probability of the predicted value is the fraction of that class within the node. For the example, the predicted Buyer class is Y (3 versus 2) and the probability is 0.6.

## Decision tree advantages

The main advantage of decision tree classifiers over other methodologies is that they are very easy to understand and interpret. The input attribute determining the first split is, with respect to the full dataset, the most discriminating of the input attributes. In general, splitting attributes at the top of the tree are more important than those lower in the tree.

Another advantage of decision trees is that in construction and in application, they do not require a lot of CPU processing power. Using decision trees to make classifications is especially fast and simple. If necessary, for a person having access to the tree structure, the classifications can be made manually in very little time.

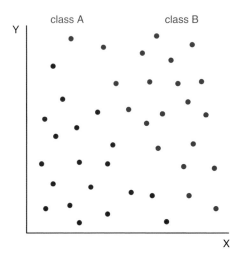

**Figure 4.3**   Classification Plot

A final advantage is that the danger of overfit by the algorithm is less than for other classification algorithms. This is especially true for training implementations that have complete split termination rules and use a validation dataset to guide check for overfitting.

### Limitations

Because decision tree algorithms consider only one input attribute at a time, they quite frequently perform worse than other classification algorithms. For example, consider the scatter plot of input attributes X and Y in Figure 4.3. A decision tree algorithm would try to split the observations based on either X or Y alone. The algorithm would not be able to locate the dividing line so visibly obvious. Other algorithms such as artificial neural networks and support vector machines would have no trouble finding and implementing the split.

Another issue is with respect to the robustness of the modeler. For example, if two input attributes are nearly equal in their discriminatory ability at the first level split, splitting with the slightly better attribute may generate a totally different tree than if the other were chosen. In fact, the overall performance of the resulting tree may be better if the second best attribute is chosen for the first split rather than the best.

## Artificial Neural Networks

**Artificial neural networks** (ANNs) are computing methodologies designed to mimic functions within the brain, giving computers the ability to learn from experience and reason. The human brain contains billions of interconnected nerve cells or **neurons**. Neurons receive inputs from other neurons via **dendrites**. Once the level of inputs to a neuron reaches a critical level, the neuron fires, sending an electrical pulse out via the **axon** to connected neurons. The connections between the output axons of one neuron and the input dendrites of a connected neuron are known as **synapses**. The synapse determines the strength of signal that passes into the connecting neuron.

Within the brain, learning is accomplished in two ways. First, the strength of the synaptic connections are altered, thus changing the strength of signal passing from one neuron to the next. Second, connections between neurons are dynamic. They may be added or removed based on experience, with the objective of improving cognitive performance. The precise mechanism for changes in synapse strengths and neuron connections is not completely understood and is still a major focus of research.

Like neurons in the brain, ANNs were designed as a network of interconnected artificial neurons that could be trained through trial and error to predict a value for the output variable, based on a set of input variable values. They are typically constructed in layers similar to Figure 4.4, depicting a neural network designed to use three input variables to predict one output. ANNs contain one or more hidden layers that are used to compute intermediate results and one final output layer where the output values (predictions) are read. The structure is known as a **feed-forward** network. The input values are fed into neurons in the first hidden layer. Each neuron uses its inputs to compute the output value transmitted to connected neurons in the next layer. The values generated by neurons in the output layer become the predictions.

All neuron input and output values are in the range [0, 1]. Therefore, the actual input values entering the network at the input nodes are first normalized

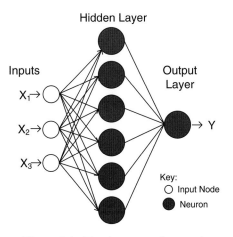

**Figure 4.4** Two layer neural network

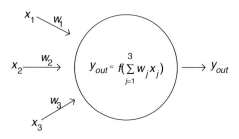

**Figure 4.5**   Neuron with Three Inputs

on a [0,1] scale. To get the actual predicted value of a network, outputs from the output layer must be denormalized.

The internal structure of an individual neuron is depicted in Figure 4.5. The output value of the neuron is the weighted sum of the inputs applied to a function $f$. Note that if the weights for a given neuron sum to one and given that all inputs are in the range [0, 1], then the weighted sum of inputs will also range [0, 1]. It is the weights, unique to each artificial neuron, that are equivalent to the synapse strengths in the human brain. The method used to determine values for these weights will be presented shortly.

The function $f$ of the neuron is known as the **activation function**. It is specified when the network is constructed and can be as simple as the identity function. An activation function frequently employed in ANN implementations is the S-shaped logistic or sigmoid function:

$$f(x) = \frac{1}{1 + e^{-a(2x-1)}}$$

where $a$ defines the steepness of the curve. See Figure 4.6. A desirable feature of this function is that inputs in the [0, 1] range generate outputs in the same range. The S-shaped curve is desirable because it somewhat mimics the "firing" of neurons in the brain. At lower input levels the output remains very low, then suddenly begins to rise as input reaches a threshold (about 0.4 in Figure 4.6).

The weights of the network are computed using the training dataset containing known values for both input and output variables. The process is as follows:

1.   Construct a network. If the ANN is to be used for regression, create just one output layer neuron. When used for classification, create one output layer neuron for each possible output value.

2.   Assign random values to each of the neuron weights.

3.   For each observation in the training set,

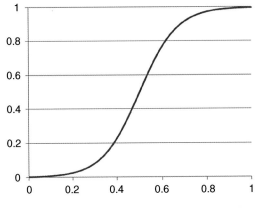

**Figure 4.6**   Sigmoid Curve ($a = 6$)

   a.  Feed the input values forward through the network to generate output value(s).

   b.  Compare the predicted output value with the actual output value.

   c.  Use a methodology known as **back propagation** to go back through the neurons, adjusting the weights slightly to improve the network's predictive performance with respect to the observation in question.

4.  Repeat step 3 until a stopping point is reached.

   Each execution of step 3 above is called an **epoch**. Depending on the size of the training set, sufficient training of the network may require thousands of epochs, while others may train in a hundred or fewer epochs. Note: The mathematics of back propagation is not presented in this text. The interested reader doing an Internet search for "artificial neural network tutorial" will find a number of excellent tutorials with in-depth explanations of back propagation.

## Overfitting the model

In Chapter 1 the concept of model overfitting was introduced. ANN models are especially prone to overfitting when care is not taken in their construction. Given enough training epochs, and sufficient neurons in the hidden layers, an ANN can be trained to almost perfectly fit any dataset. Consider the points in Figure 4.7a. A good regression model for the data is represented by the curve in Figure 4.7b. Yet, an overtrained ANN regression could produce a model similar to Figure 4.7c. Which model, b or c, would you expect to generalize better?

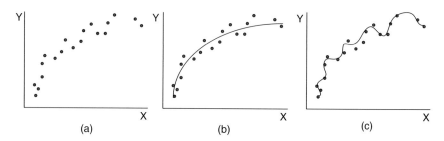

**Figure 4.7**   Regression Model Overfitting

To avoid overfitting, the training process needs to be stopped before reaching an overfit state. For example, when training the dataset of Figure 4.7a, the model would very likely progress through a state similar to Figure 4.7b before reaching the state of Figure 4.7c. The trick then is to recognize when the model has reached a point similar to Figure 4.7b and stop. Note that the value of $r^2$ for Figure 4.7c, if computed, would be significantly better than that for Figure 4.7b.

One way to recognize the point at which overfitting begins is to first split the dataset into a subset of training observations and a subset of validation observations. Start the model building process using the training dataset. At regular intervals, as the training progresses, pause to compute a pair of model performance measures – one using the training set and the other based on the validation set. For example, in a classification model compute the error rate. Compute the measure once using the training set and a second time using the validation set. When the training first starts, you are likely to see the performance measure improve for both datasets. However, as training progresses you may see the performance measure improve with respect to the training dataset, while the validation measure gets worse. When this happens, the model is very likely to be overfit.

## Moving beyond local optima

One of the problems encountered in the development of early artificial neural networks was that of local optima. The initial neuron weights are random values. As the neural network is training, it takes small steps (weight adjustments) in a direction that will produce the best improvement. These steps may move it toward a solution that is not necessarily the best possible. To illustrate, consider the maps of Figure 4.8. Although this is not a perfect representation of an ANN search space, it sufficiently depicts the process.

Think of Figure 4.8a as representing the search space of ANN weights where the darker areas represent regions of better performance. The optimal location is marked "Best". The random starting position is marked by the "X". At each

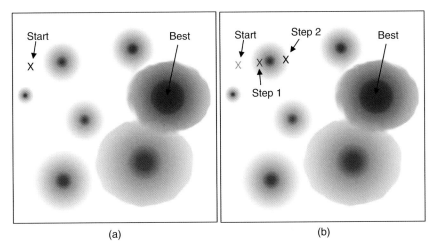

(a)                                        (b)

**Figure 4.8**   ANN Optimal Surface Map

training iteration, the current location takes a step in the direction of greatest improvement. Figure 4.8b represents the state after two iterations. At step 1, the ANN moved toward the nearest local optimum. At step 2, it continued in the same direction, which pushed it just past that same local optimum. Where will it go at step 3? It will move back toward the nearest local optimum, which is back toward where it was at step 2. Without changes to the algorithm, it will continue to vacillate around the local optimum never approaching the optimal ("best") location.

To overcome this problem, researchers and designers of ANN algorithms added two adjusting parameters to the backward propagation methodology. The first added was a **learning rate**. Think of the learning rate as the distance moved at each iteration. With a higher learning rate, it is possible that the ANN will step far enough beyond a local optimum to move, at the next iteration, toward a different optimum.

The second parameter added was **momentum**. The momentum attempts to keep the ANN moving in the same direction as movement at the last iteration. Momentum ranges in value between 0 and 1. It represents the portion of the movement from the previous iteration that is added to the current iteration's computed movement to produce a total momentum adjusted movement. When momentum is zero, only the movement of the current iteration is used (Figure 4.9a); when momentum is one, the full movement of the previous iteration is included (Figure 4.9b). At 0.5, for example, half of the movement of the previous direction is added to the current movement, thus nudging the training in the direction of previous movement, while allowing the current iteration to contribute its computed direction of movement (Figure 4.9c).

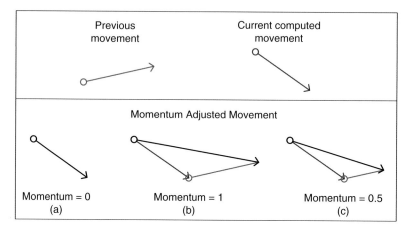

**Figure 4.9**   Momentum Adjustments

The addition of learning rate and momentum parameters has improved the ability of ANNs to train toward more globally optimal locations. The question then becomes, "What should the learning rate and momentum be set to, in order to optimize the training process?" The answer varies from dataset to dataset. There is no single best combination of parameter settings. VisMiner's method for setting these two parameters, while training, is presented in Chapter 5.

## ANN Advantages and limitations

As previously pointed out, the primary advantage of artificial neural networks is the ability to fit almost any dataset. Yet this is also a weakness – if not monitored carefully during training, it is very easy to overfit the model to the training data. The VisMiner ANN monitoring process, which will assist in avoiding an overfit model, is presented in Chapter 5.

Other issues with respect to ANN models are:

- Reproducibility of results – the initial neuron weights of the ANN are random values. Therefore, depending on the initial starting position, you are not guaranteed to arrive at the same final solution from run to run and, as previously stated, that solution may not be an optimal solution. With most datasets, this is not usually an issue if the learning rate and momentum parameters are monitored and adjusted as training progresses.

- Interpretability of the model – from a mathematical perspective, ANNs are complex. You cannot look at a single coefficient value in the model formula to determine the contribution to the overall prediction of a single input attribute.

- Tests for model significance – at this time, there is no published test of model significance. By significance, we mean: "Can we measure the probability that the model predictions are anything more than chance occurrences?" Unlike the well-studied linear regression models, there is no "null hypothesis" type test. When using ANNs, the only work-around is to build the model with enough observations that we feel comfortable that the model actually will predict accurately when applied to new data. The use of validation data, discussed in Chapter 5, can give us that assurance.

## Support Vector Machines

The algorithm for **support vector machines** (SVM) was developed to improve classifier performance. Look back at the scatter plot of Figure 4.3. If a decision tree algorithm is applied to this dataset, it will perform poorly. This is because the algorithm considers input attributes in isolation when searching for the best split attribute. As a result, the splits are always parallel to one of the attribute axes. On the other hand, in its simplest form, the SVM computationally locates that line that will best split the observations according to classification attribute values.

In the search for that dividing line, there are a number of issues that the algorithm needs to deal with. Consider Figure 4.10 – a slightly modified version of Figure 4.3. Three candidate dividing lines are drawn separating the red dots (class A) from the blue dots (class B). Each would work equally well as a

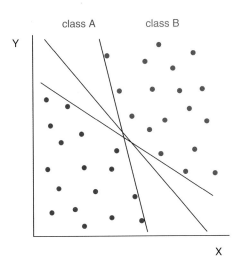

**Figure 4.10**   Potential Dividing Lines

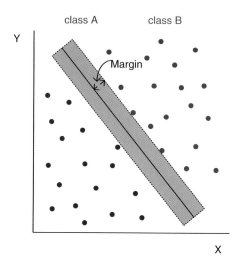

**Figure 4.11**    Dividing Line Margin

classifier. Is one better than the other? Intuitively, one might expect that the center line would generalize better. How might this center line be computationally located?

Suppose that now, instead of using a line to divide the points, we use a bar (Figure 4.11). If that bar is rotated as needed, forcing it to be as wide as possible without overlapping any of the points, the length-wise bisecting line will be that dividing line expected to generalize best. (Note: For the interested reader, proofs that this is indeed the dividing line that will generalize best are found in numerous papers and books on support vector machines.)

One-half of the width of the bar in Figure 4.11 is known as the **margin**, which is interpreted as the distance the dividing line can be moved without introducing classification error. Hence, the classifier construction problem can be recast as that of locating the dividing line having the greatest margin.

Consider now the problem where it is not possible to totally separate the points with a dividing line (Figure 4.12). The solution is to add a slack value ($z$) to each of the offending points that would be equal to the amount needed to push it back to the non-offending side of the margin, while recognizing that for most points in the dataset, the required slack is zero. To accomplish this, the dividing line needs to be chosen such that it minimizes the total slack requirements. The problem now becomes one of maximizing the margin while at the same time minimizing the total slack requirements. Although in this text, we do not go into the mathematics of the search mechanism for locating the optimal solution, suffice it to say that it uses a combination of the two objectives – maximizing the margin and minimizing the total slack.

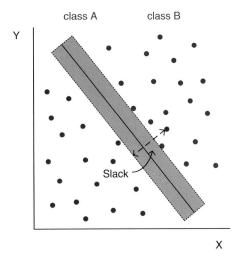

**Figure 4.12**  Slack Requirement

During the search, to balance the contribution of the two objectives, the slack total is adjusted by a cost factor (*C*). Such that:

$$slack\ contribution = C \sum_{i=1}^{m} z_i$$

where

$z_i$ is the slack requirement of the $i^{th}$ observation, recognizing that most will be zero

*m* is the number of observations.

The value of *C* is set by the analyst prior to algorithm execution.

## Data transformations

Look at Figure 4.13a. Attempting to find a dividing line for this dataset would be futile. However, suppose that instead of submitting (X, Y) observation pairs to the algorithm, we submit (X², Y²), as plotted in Figure 4.13b. The SVM will have no trouble finding a dividing line for this set. The point here is that a transformation of the data prior to processing by the algorithm may generate classification results not achievable by the original data.

Most implementations of SVM algorithms include a set of potential transformations from which the analyst may choose. These are known as

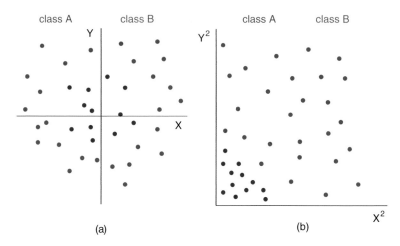

**Figure 4.13**   Non-Linear Dividing Line

the **kernel** functions. Typical functions implemented are: linear, polynomial, radial basis, and sigmoid. By default, most SVM algorithms use the radial basis function. Finding the function that performs best is, for the most part, a trial and error process.

## Moving beyond two-dimensional predictors

In the real world, most classification models will be built using more than two predictors. To facilitate visual understanding, the dataset plots of Figures 4.10–4.13 contained just two predictors. However, moving beyond two input dimensions using SVMs is not a problem. In three dimensions, instead of a dividing line, a dividing plane is located. In four or more dimensions, it becomes a hyperplane. Even though it is not possible to visually represent a hyperplane, the mechanics of the algorithm are still the same.

## SVM Advantages and limitations

Like the ANN, SVM classifiers can be made to fit almost any dataset, depending on the chosen kernel function and the inputs selected. This is a two-edged sword. A good fit is important, yet like the ANN, SVM classifiers frequently overfit the data.

The best way to avoid this problem is to build multiple SVM models using varying combinations of kernel functions and cost parameter settings, then comparing results against a validation dataset and choosing the combination with the best fit to the validation set.

Currently in VisMiner, the SVM modeler is a black box. It does not support analyst tweaking of the parameters. By default, VisMiner uses the radial basis function. During model construction, with a small subset of the data, it builds multiple models using varying values of the cost parameter, choosing the best, then applying to the full dataset. The analyst only sees this final model.

In choosing to build an SVM classifier, another consideration is that of processing time. SVM algorithms are quite CPU intensive. Typically they will take much longer to complete than decision tree or ANN processors. You may want to build using smaller datasets or fewer predictor variables.

## Summary

VisMiner implements algorithms for both classification and regression. These modelers can be deployed by the analyst without having knowledge of the algorithm internals. However, having that model understanding can result in better model application decisions and less ambiguous interpretation of the results.

A problem common to all prediction modelers is the overfitting of the model to the training data. Each modeler has guidelines and processes that can be followed to avoid this problem. A knowledge of the algorithm internals helps in their application.

A summary of the advantages and limitation of each algorithm is found in Appendix A, Table A.1.

# 5

# Classification Models in VisMiner

Classification is a form of prediction modeling that uses selected input attribute values to predict a nominal or categorical output value. In constructing a classification model, a dataset is used that contains historical data from past events in which the values of both the input and output attributes are known. The classification methodology uses those values to construct a model that best fits the data – that is the model accurately predicts the output category based on input values. The process of model construction is sometimes referred to as **training**. Once constructed and validated, the model can be used in the future to predict the category when the input attribute values are known, but the value of the output attribute is not yet known. For example, an insurance company may want to build a classification model to predict if an insurance claim is likely to be fraudulent or legitimate.

This chapter introduces the functionality of three modelers or methodologies for classification as they are implemented in VisMiner: decision trees, artificial neural networks, and support vector machines.

## Dataset Preparation

The dataset used for classification modeling in VisMiner must be in a tabular format. The input attributes may be of any data type – numeric, ordinal, or nominal. The output attribute must be nominal or discrete (integer) numeric.

It is important to remember that when using VisMiner to build a classification model, the dataset should contain only the attributes (input and output) to be used by the modeling process. There should not be any row identifiers or other attributes included. Therefore, if your dataset contains unneeded attributes, before starting the modeler, create a derived set containing only the attributes that you want to include. (See Chapters 2 and 3 for details on how this is done.)

*Visual Data Mining: The VisMiner Approach*, First Edition. Russell K. Anderson.
© 2013 John Wiley & Sons, Ltd. Published 2013 by John Wiley & Sons, Ltd.

In preparing datasets for classification, consider also the cardinality of nominal attributes. Normally for classification modeling, VisMiner supports attributes with a maximum cardinality of 10. The only exception is for decision trees where the maximum is bumped up to 30. Any attempts by the user to create models from datasets with cardinalities above the maximum allowed will be rejected.

## Tutorial – Building and Evaluating Classification Models

We begin with a simple dataset – Iris.csv which was explored in Chapter 2. Our objective is to build a model that correctly classifies iris varieties based on the four flower measurements: petal width, petal length, sepal width, and sepal length.

☞   In the VisMiner Control Center, open the Iris.csv dataset.

☞   In VisMiner, modeling algorithms are applied by dragging the desired modeler implementing the algorithm down over the target dataset. To start, drag the "Dec Tree Classifier" down to the Iris dataset and release.

As the modeler processes the dataset, the gears of the modeler turn. You know that it has finished when the gears stop and then disappear. Because the dataset is so small and the decision tree algorithm is quite simple, the modeler will finish almost immediately. When you process large datasets, depending on the algorithm, it may take minutes or sometimes even hours to complete. While the modeler is processing, the Control Center is still active. You may perform other operations while you wait, such as preparation of different datasets or exploration using the data viewers.

In the case of the Iris dataset, the classification modeler immediately begins processing as soon as the modeler is released over the dataset. This is because there is only one nominal attribute in the dataset (variety). In datasets where there are multiple nominal attributes, you will first be required to identify the output (classification) attribute before processing begins.

## Model Evaluation

After processing is complete, it is time to evaluate the resulting model. There are two objectives of the evaluation:

• How well does the model perform?
• How do inputs contribute to model predictions?

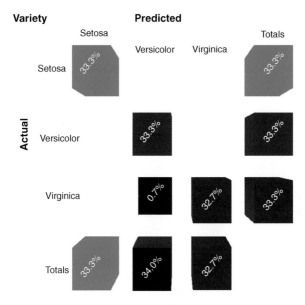

**Figure 5.1**   Confusion Matrix – Iris data

Beginning with the first objective,

⇪ Drag the model up to an available display and release.

⇪ Select "Confusion Matrix".

The result is in Figure 5.1. In the title bar (top left) is the error rate. For this classifier it is 0.7% – indicating that out of the 150 observations only one was misclassified. The standard confusion matrix represents a N × N array of observation counts or ratios where N is the cardinality of the output attribute. The N rows represent the actual values in the data; the N columns represent the predicted values. Correct classifications are represented by those down the main diagonal. The VisMiner confusion matrix is plotted in 3-D and can be rotated in the same way as other VisMiner visualizations. The Z axis represents the count in each cell. The percentages are with respect to the total observations. The colors serve to distinguish between possible predicted/actual combinations. The error rate is the sum of all cells not on the main diagonal.

From the confusion matrix we see that the only misclassification was that a single Virginica observation classified as Versicolor. Based on the extremely low error rate, our classifier appears to do a good job.

When building classification models, it is a good idea to construct multiple models using different classifiers, then compare the results. VisMiner supports two other classification modelers: support vector machine (SVM) and artificial neural network (ANN).

↺ Close the confusion matrix for the decision tree classifier.

↺ Drag the SVM classifier (support vector machine) to the Iris dataset.

↺ After processing is complete, hover over each of the two models to view and compare error rates.

At this point, we shift focus to ANN model construction. We will get back to the decision tree and SVM models once the ANN has been built.

The VisMiner ANN classifier implements a feature that allows the user to interact with the algorithm during the model building process. For this reason, its execution is a little more complex than that of the simpler drag and drop of the decision tree and SVM modelers.

↺ Drag the ANN classifier to the Iris dataset.

↺ Select "Build interactively".

The interactive build option supports user control over the training process (Figure 5.2). The effectiveness of ANN training depends on the network learning rate and momentum, yet there is no single best learning rate or momentum for all datasets. It will vary from dataset to dataset. The interactive build option allows you to monitor the training progress while adjusting the learning rate and momentum as it progresses.

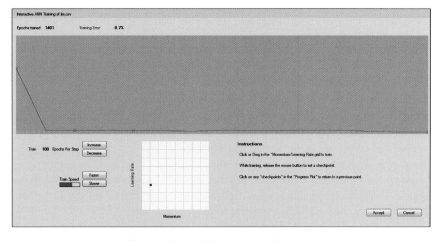

**Figure 5.2**   ANN Interactive Training

The interactive build begins after the automatic completion of a single epoch pass using the learning rate and momentum represented by the red dot in the grid – a very high learning rate paired with a low momentum. The initial "Training Error" is shown above the "Training Progress Plot". Normally the initial error will be high as it is dependent on the random weights assigned to begin the process. It will improve as training progresses.

Each time you press the mouse button down while over the grid, training resumes using the learning rate and momentum corresponding to the current mouse location. It continues until the button is released. While training, you may drag the red dot to other locations within the grid to change the learning rate and momentum on the fly. The progress plot is updated to show the current training error. When the mouse button is released, a checkpoint is created. Checkpoints record the current state of the ANN and are depicted on the progress plot by small red circles. They allow you to go back to previous checkpoints to try training in a different direction. The interactive process allows you to visually search for the right combination of learning rate and momentum applicable to the dataset.

☞ While holding the mouse down, slowly drag the red dot down from the upper left corner toward the bottom center.

As you drag, you should see the training error progress plot begin to drop. While training, it is a good idea to release then press the mouse button every few seconds to set checkpoints that you may want to return to. Note: There will always be an initial checkpoint available which is set immediately after completing the first training epoch.

As you train, keep in mind the error rates achieved by the decision tree and SVM classifiers (2.7% and 2.0% respectively). If you cannot push the ANN error below these values, it probably is not a useful model.

☞ Continue the training process until the classification error flattens (0.7%).

☞ Click "Accept".

After processing has completed, drag each new model up to an available display, selecting "Confusion Matrix" as the viewer.

How do these two new classifiers (SVM and ANN) compare in terms of error rate with the decision tree? The ANN misclassified only one flower (0.7%), while the SVM and decision tree misclassified three (2.0%) and four (2.7%) respectively. With just 150 observations, it is not possible to definitively say that the ANN Iris classifier is better than SVM or decision tree classifiers. Later in the chapter we will revisit issues of model performance.

☞ Close all open confusion matrices.

Variety: Setosa Versicolor Virginica

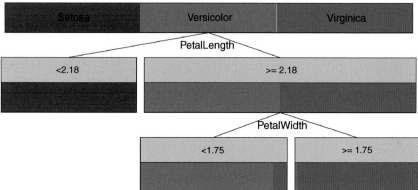

**Figure 5.3**   Tree Graph – Iris data

Let's look at the second objective of our evaluation – understanding input contribution.

Drag the decision tree model up to an available display and select "Tree Graph".

In the tree (see Figure 5.3) all nodes are sized and color encoded according to the contents of the node. The top bar in the tree represents the entire dataset. All varieties are equally represented given that the red, blue, and green bars are of equal length. In processing, the decision tree classifier chose *PetalLength* as the most discriminating attribute in making the first split. The split cut-off was set at 2.18 centimeters. The second level in the tree depicts this split. All Setosa observations went into the left node, while the right node contains both the Versicolor and Virginica. Look back at Figure 2.14 to understand the rationale for the chosen split. The next split based on *PetalWidth* did not generate any totally homogeneous nodes and thus required additional splitting.

As the tree progresses toward the leaf nodes, at times there is not enough room to show the split criteria. In these cases, the criteria box either draws a " . . . " or is left blank.

☞   To see the split criteria and node contents, hover over any of the nodes.

The leaf nodes of the tree all contain homogeneous content, except for the node containing two Versicolor observations and one Virginica observation. This node represents the one error in the confusion matrix. When the decision tree is used to make a prediction, the input attribute values are used to navigate to a leaf node. The most frequently occurring category of that leaf node then becomes the predicted value.

## Exercise 5.1

Use the OliveOil dataset to generate classification models based on the acid measures.

a.   Build classification models to predict Region using the decision tree, ANN, and SVM classification modelers. Note: The modelers automatically use all attributes in the dataset for model construction. Since you do not want to use Area to classify Region, you will first need to create a derived set that excludes Area, then build the models using the derived set. Look at the confusion matrices for all three models. How well do they predict the training set values?

b.   Build classification models to predict Area using the decision tree, ANN, and SVM classification modelers. Look at the confusion matrices. How well do they predict the training set values? Which modeler performs best? Which Areas do the models have the most trouble predicting? Hint: The cells in the matrix off the main diagonal (excluding the totals column and row) with the tallest bars represent the observations most frequently misclassified.

c.   View the tree graph for the decision tree model. Which acid best distinguishes the South Apulia oils? Describe the primary distinguishing acids characteristics of the Inland Sardinia oils.

## Prediction Likelihoods

To this point, we have only evaluated input contributions of the decision tree models using the tree graph. Decision trees are relatively simple structures. The structure of other models is not as easy to visualize due to the complexity

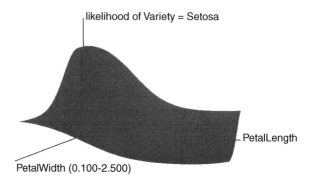

likelihood of Variety = Setosa

PetalLength

PetalWidth (0.100-2.500)

**Figure 5.4**   Classification Surface – Iris data

of the algorithms. The "Classification Model" viewer was designed to support this task.

⤵ Drag the SVM model for the Iris dataset up to an available display, release and select "Classification Surface".

The visualization that opens (Figure 5.4) shows an XYZ surface plot of the likelihoods of the selected output attribute value given the input attributes. The likelihoods, as plotted on the Y axis, range in value between zero and one. As with other 3-D viewers in VisMiner, the plot may be rotated to gain a better perspective of the actual surface shape. Scales on the axes are omitted in order to focus the user's attention on the surface shape rather than point reading.

In the plot you can see that the likelihood of being variety Setosa is greatest for low values of PetalLength and PetalWidth. As both the length and width increase the likelihood drops.

Notice that the point of greatest likelihood is only about two-thirds of the way up the Y axis – close to about 0.7 likelihood. You might be wondering if there aren't combinations of PetalWidth and PetalLength that have greater likelihoods. The answer can be found in the option panel to the right (Figure 5.5). There are four sliders – one for each of the input attributes that were used to build the model. The top two, for PetalLength and PetalWidth are disabled, because they are currently selected on the X and Z axes respectively. The slider for SepalLength is set at 6.1 and the slider for SepalWidth is set at 3.2. The surface plot that you see represents possible combinations of PetalLength and PetalWidth when SepalLength is fixed at 6.1 and SepalWidth is fixed at 3.2. Since SepalLength and SepalWidth were used to build the model, they also contribute to the output likelihoods.

⤵ Drag the slider for SepalLength to the left.

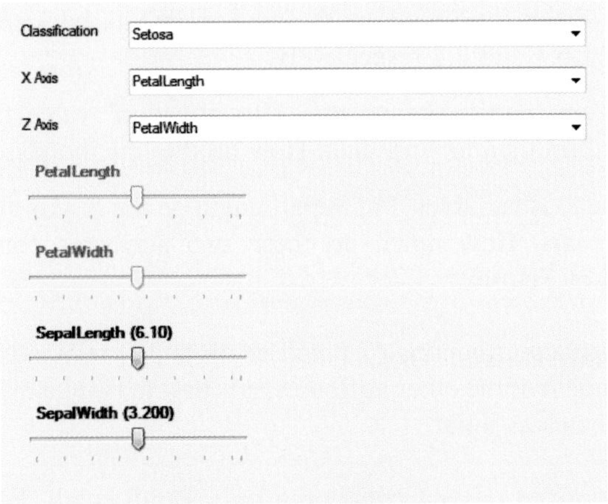

**Figure 5.5**   Option Panel

As you move to the left, you see the surface plot move upwards, indicating that as *SepalLength* decreases, the likelihood of the Setosa variety increases.

⏚  Drag the SepalLength slider to the right.

As you drag, the likelihood drops – iris flowers with long sepals are very unlikely to be of the Setosa variety.

⏚  Drag the SepalLength slider back to 6.1.

⏚  Drag the SepalWidth slider to the left and to the right.

The contribution of SepalWidth to the likelihood of being Setosa is opposite that of SepalLength. That is, as SepalWidth increases the likelihood of being Setosa increases.

In addition to the up and down movement of the plot surface, as you changed the values of SepalLength and SepalWidth, did you notice a change in shape of the plot surface? When you see a change in shape, it is indicative of **interaction** between the axis attributes and the attribute being changed. Interaction means that the nature of the contribution of one attribute changes depending on the level of a second attribute. That is, it does not make the same contribution independent of the other input values. When you see the plot surface rise or fall without a dramatic change in shape, the contributions of each are said to be **independent**.

↪ Drag the sliders for SepalLength and SepalWidth back to their original positions at 6.1 and 3.2 respectively.

↪ In the options panel, change the "Classification" value from Setosa to Versicolor and then to Virginica. How do the likelihoods compare?

↪ Change the axis variables. Put SepalLength on the X axis and SepalWidth on the Z axis. How much do these two attributes contribute to the likelihood of Setosa?

To better grasp contributions of PetalLength and PetalWidth in classifying Variety independent of the other attributes, it is best to build a model using only those two attributes as input.

↪ Create a derived dataset containing only PetalLength, PetalWidth, and Variety.

↪ Drag the SVM Classifier down to the newly derived dataset to build another model.

↪ View the confusion matrix for this model. How does it compare with the model that used all four input attributes?

↪ View the model in the Classification Surface viewer side-by-side with the original four input SVM model.

The resulting surfaces for Setosa, Versicolor, and Virginica are in Figure 5.6. The likelihood of Setosa is greatest when both PetalLength and PetalWidth are low. The likelihood of Versicolor is greatest when both are in the mid-range. They drop to much lower values when only one of the two is in that mid-range. The likelihood of Virginica is greatest when the values of both are high.

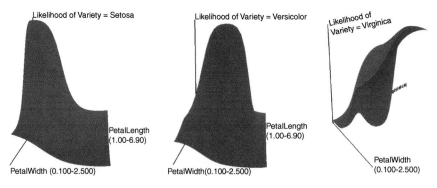

**Figure 5.6** Likelihood Plots – Setosa, Versicolor and Virginica

# Classification Model Performance

Earlier in the chapter, classification model performance was judged by simply looking at the model error rate. You are now ready for a more in-depth evaluation of model performance.

➪  Open the dataset CarBuyer. See Table 5.1 for a description of attributes.

➪  Review the summary statistics

The CarBuyer dataset contains demographic data purchased from the Tell All insurance company by Wheels, a boutique sports cars manufacturer. The dataset was used to identify prospective customers from an earlier promotion. Wheels

**Table 5.1**  CarBuyer Attributes

| Attribute Name | Description |
| --- | --- |
| Age | age of individual |
| BicycleOwner | Y – owns a bicycle; N – does not own bicycle |
| BoatOwner | Y – owns a boat; N – does not own boat |
| Buyer | Y – purchased sports car in previous promotion; N – did not purchase |
| CommuteDistance | distance in miles that individual commutes to work |
| CreditRating | individual's credit rating (0–10) |
| EducationYrs | number of years of formal education |
| Gender | F – female; M – male |
| Height | height in inches |
| HomeLocation | R – rural; S – suburban; U – urban |
| Income | annual income in dollars |
| JobCategory | clerical, managerial, professional, sales, skilled manual, unskilled manual, retired, or unemployed |
| JobYrs | years at current job |
| MaritalStatus | D – divorced; M – married; S – single; W – widowed |
| MotorcycleOwner | Y – owns a motorcycle; N – does not own motorcycle |
| NbrChildren | number of children living at home |
| NbrPassCars | number of passenger cars currently owned |
| NbrPickups | number of pickups currently owned |
| NbrSpeedingCitations | number of citations for speeding received in last three years |
| NbrSportsCars | number of sports cars currently owned |
| NbrSUVs | number of sport utility vehicles currently owned |
| Region | East, Midwest, Pacific, South, or West |
| RVOwner | Y – owns a recreation vehicle; N – does not own a recreation vehicle |
| SecondHomeOwner | Y – owns a second home; N – does not own a second home |
| Weight | weight in pounds |

merged the demographic data from Tell All with the actual purchase records from the promotion. They are now considering another promotion using more data from Tell All. Wheels' management would like to know the feasibility of repeating that last promotion. This time, instead of offering the promotion to all individuals on the Tell All list, they would like to use the historical dataset to build a classification model that will identify the best candidates for the promotion.

Wheels' plans are as follows:

1.  Purchase a 20,000 person dataset from Tell All at a cost of $10 per person.

2.  Prepare a comprehensive, glitzy brochure about the Wheels sports car. The preparation costs are estimated to be $15,000. The printing and mailing costs are estimated to be an additional $12 per brochure.

3.  Send the brochure to the identified candidates offering a one week, no strings attached, free trial of the car. The one-week trial is expected to cost Wheels $210 per trial. The offer is too good to pass up. All recipients of the offer are expected to accept.

4.  Wheels estimates that the marginal contribution to profit of each car sale is $900.

The problem therefore, is to build a classification model, then use that model to estimate the overall expected profit or loss of the promotion.

The first step in building the model is to select the input attributes (variables). The output attribute is Buyer. When selecting the inputs, it is good to remember that models built with fewer input attributes are more likely to generalize better than those built with more.

☞   In the summary statistics hover over the cardinality for Buyer. Notice that buyers make up just over 8% of the total population (1,648 of 20,000).

☞   Open the dataset in the correlation matrix.

☞   Open a scatter plot for synchronization with the correlation matrix.

☞   To better understand the data, look at correlations between potential attributes. Are there any surprises? Do the correlations appear reasonable and expected? For example, the correlation between HomeLocation (R-rural, S-suburban, and U-urban) and CommuteDistance is one of the strongest. Click on that cell in the correlation matrix to bring it up in the scatter plot (Figure 5.7). The plot shows that suburban and rural residents have about the same mean commute distance, although the spread of the rural residents is greater. Urban residents overall have much shorter commutes. Look at the strongest inverse correlation – Income versus CreditRating. Does this surprise you? Looking at the plot (Figure 5.8), you see a lot of low income individuals with high credit ratings. This explains the inverse correlation.

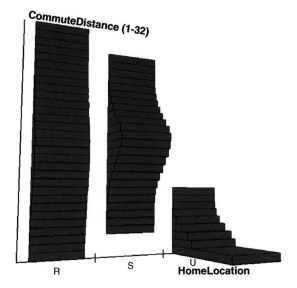

**Figure 5.7**   Scatter Plot HomeLocation vs CommuteDistance

**Figure 5.8**   Scatter Plot Income vs CreditRating

☞ Look over the cells on the Buyer column and row. Color-wise, none of the potential inputs stands out as highly correlated with Buyer. You will need to hover over each individually to see the correlations.

☞ Remove the attributes that do not appear to have any significant correlation with Buyer by clicking on the attribute name. For purposes of continuity in this tutorial, remove: BicycleOwner, BoatOwner, CommuteDistance, EducationYrs, Height, HomeValueIndex, JobYrs, MotorcycleOwner, NbrChildren, NbrPassCars, NbrPickups, NbrSUVs, RVOwner, SecondHomeOwner, and Weight.

☞ Create a subset of the remaining attributes. Name the subset "Selected".

☞ Close any open viewers.

In order to allow the modelers to focus on attribute relationships that best discriminate between the car buyer and the non-buyer, it sometimes helps to balance the ratio of buyers to non-buyers.

☞ View the Selected dataset in the parallel plot.

☞ Use the filter sliders to select only buyers (Buyer = Y).

☞ Right-click on a slider to make a new dataset. Name it "Yes".

☞ Adjust the same filter sliders to select only non-buyers (Buyer = N).

☞ Right-click on a slider to make a new dataset. Name it "No".

☞ Close the parallel plot.

☞ In the Control Center, create a derived dataset from the newly created non-buyer dataset (No). The derived set should contain all of the attributes (columns) and 1,648 rows (the same number as the Yes set). Name the set "reducedNo".

☞ Drag the reducedNo dataset over the Yes dataset, release, and merge. Your dataset panel should look similar to Figure 5.9.

You now have a dataset (Yes-reducedNo) consisting of the selected attributes from the original carBuyer dataset and containing 3,296 rows – half of which are car buyers and half are non-buyers. Before modeling, the dataset needs to be split into training data (used to construct the model) and validation data (used to verify that it will generalize).

☞ Create a derived dataset from the Yes-reducedNo dataset. Select all of the columns. Use 25% (824 rows) for validation, leaving 2,472 for training. Name the set "Target" – implying that this is the fully prepared target dataset to be used for modeling.

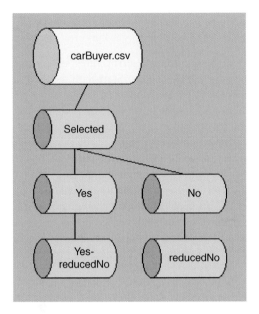

**Figure 5.9**   Derived Datasets

As you work with this dataset in the instructions that follow, your results will vary slightly from those shown in the examples. The split of rows into training and validation sets is a random process. Hence, your split will not exactly match the split used to prepare the examples.

⤶   Create two classifiers by dragging the "Dec Tree Classifier" and "SVM Classifier" modelers down to the Target dataset. Select Buyer as the classification column.

⤶   View the confusion matrix for each model after it has finished processing. The decision tree should finish first.

You do not have to wait for the SVM to finish before opening the decision tree's confusion matrix. How well does each perform? In the Target dataset created, the rates for the decision tree classifier were 11.3% (training) and 14.1% (validation). Because there is not a large difference between the training and validation rates, we can conclude that it is not overfit; it generalizes well. The SVM classifier is quite different. The error rate for the training data is an almost perfect 1.0%, but soars to 16.5% in the validation data. Clearly the SVM model is overfit.

⤶   Now that you have seen the respective results for the decision tree and SVM classifiers, interactively create an ANN classifier using the Target data. Try to get its validation error rate below that of the other two. Based on previous experience, you should be able to get the error rate down to about 11%.

Judging classifiers by comparing error rates is a simple yet superficial approach. Let's now dig a little deeper. Start with the ANN classifier: if based on error rate, it appears to perform best.

In making predictions, classification models compute the probabilities that each possible output class value is the actual value. It then chooses the class value with the greatest probability. For example, the CarBuyer classifier, for a given observation, may generate a Buyer probability of 0.84 for "Y" and 0.16 for "N". Based on the probabilities, the predicted class is "Y". Another observation may generate a Buyer probability of 0.505 for "Y" and 0.495 for "N". Again, the predicted class is "Y". Your confidence in the accuracy of the first prediction should be much greater than in the second.

The confusion matrix computes its error rates based on the predicted values only. It does not take into account differences in probabilities when counting those predictions.

For example, if a classifier is used by Wheels to determine which potential customers should or should not receive the planned promotion offer, there are two errors possible; individuals that would be buyers, classified as non-buyers (false negatives); individuals that would not buy classified as buyers (false positives). In the first case, Wheels misses out on a sale because the offer is not made. In the second, Wheels makes promotional expenses ($12 + $210) on an individual that does not buy.

Because the second error actually involves wasted out-of-pocket expenses, suppose that Wheels would like to eliminate as much as possible this error. It can do this by adjusting the cut-off. By default in a binary decision, a prediction of carBuyer = N is made if the probability of N exceeds 0.50. Suppose that instead of setting the cut-off at 0.50 it is dropped to 0.30. More individuals will be classified as non-buyers because the bar is set lower. Consequently fewer individuals will be classified as buyers (only those with carBuyer = Y probability over 0.70). However, in raising the bar, Wheels should have more confidence that those individuals actually receiving the promotion will buy. At the same time, Wheels must also recognize that the number of individuals that would have bought but did not receive the promotional offer will increase.

While viewing the confusion matrices of the car buyer classifiers, you probably noticed the cut-off box (Figure 5.10) in the panel above the matrix

**Figure 5.10**   Confusion Matrix Probability Cut-off

plot. By dragging the green slider to the left or right the user is able to adjust the probability cut-off at which a prediction is triggered.

⯈ Drag the cut-off slider for the confusion matrix of the tree classifier down to 30%.

As the cut-off for Prob(N) is dragged to the left, the false positives (wasted promotions) begin to drop while the false negatives (missed sales) increase. At this point, you might ask, "At what level should the cut-off be set to eliminate all false positives?" Or you might ask, "At what level should the cut-off be set to eliminate all false negatives?"

⯈ Drag the cut-off slider to determine if and at what level the false positives disappear and at what level the false negatives disappear.

## Interpreting the ROC Curve

To dig deeper into the performance of the ANN car buyer classifier:

⯈ Drag the model up to the display currently containing the confusion matrix and release on the right side, moving the confusion matrix to the left.

⯈ Select "ROC Curve" (Figure 5.11).

The ROC (**receiver operating characteristic**) curve represents the trade-offs between the undesirable **false positive rate** (FPR) and the desirable **true positive rate** (TPR). That is, how much of an increase in the FPR must be accepted in order to achieve a desired increase in the TPR?

The ROC relies on the fact that the confidence we have in predictions varies from observation to observation. Suppose that, instead of predicting a positive result if the probability of the positive value exceeds the probability of a negative value, we decide to only predict positive if the probability is 0.90 or greater. Thus we would expect the FPR to be low (0.10). But what would the TPR be? If 75% of the positive observations have a positive probability of 0.90 or greater, then the TPR will be high. However, if our classifier is less certain and only 25% of the positive observations have a probability of 0.90 or greater, then TPR will be much lower.

The ROC visually represents how confident the classifier is in its predictions. The ROC of an ideal classifier would go straight up the left axis to the top, then horizontally across the top to the right, indicating that it can achieve a TPR of 1.0 without including any false positives (FPR = 0).

The closer that a ROC curve is to the ideal, the more confidence we have in its predictions. A common measure of closeness to the ideal is the **area under the**

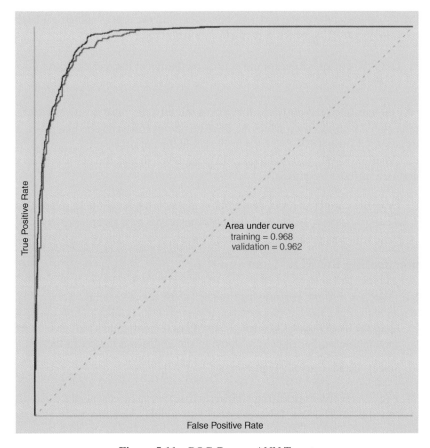

**Figure 5.11**   ROC Curve – ANN Target

**curve** (AUC). Given that FPR and TPR both range between 0 and 1, the AUC of the ideal classifier is 1.0. The dotted line in the ROC viewer represents the ROC of a random guess. Its AUC is 0.5. Hence, for a classifier to be considered useful in any way, its AUC should be greater than 0.5.

☞ Indicate to the ROC viewer which class value you define as the positive result by selecting "Y" in the "Positive result" drop-down.

The ROC viewer presents two curves – one computed using the training data and the other using the validation data. The AUC of both curves is high (above 0.96 in Figure 5.11). Given that the classification modeler did not see the validation data until after the model was complete, the validation ROC curve should normally be used when judging the performance of classifiers and in making comparisons between classifiers.

☝ Compare the decision tree and SVM classifiers to the ANN classifier by opening ROC viewers for each. Which is best?

☝ Close all ROC viewers except for the ANN viewer.

When using a classifier to selectively identify cases for treatment, one is not required – nor is it necessarily wise – to treat all predicted positive responses. For example, an insurance company may use a classifier to predict if an uninvestigated claim is "fraudulent" or "legitimate". If predicted as "fraudulent", then the insurance company will spend money conducting a thorough investigation of the claim. If predicted as "legitimate", the insurance company will pay the claim without additional investigation. As in all binary classification problems, there are the two possible error types: **false positives** and **false negatives**. With respect to the insurance company, a false positive would have them spending additional money, perhaps $5,000, to investigate a claim that is actually "legitimate" and should be paid. A false negative would have them paying off a "fraudulent" claim, perhaps $250,000. Because the cost of a false negative is significantly higher than that of a false positive, the insurance company may set the bar lower and investigate all claims in which the probability of "fraudulent" is at a predetermined level lower than 0.50 – say, 0.30, for example.

In the case of Wheels, a false positive will cost $222 ($12 to send the brochure plus $210 to loan the car) to court a customer that will not buy. There is no cost for a false negative. Wheels will only have missed out on the opportunity to sell a car to someone that would have bought.

The **Lift Chart** was designed to assist in assessing a classifier's ability to select only the most likely positives.

☝ In the "Curve type" drop-down of the ROC viewer, select "Lift Chart" (Figure 5.12).

The X axis of the chart represents the top X% of the observations to be selected. It is based on a sort of the observations by probability of being positive in descending sequence. Hence, a cut-off value of 25% would mean that we are selecting the top 25% most likely observations to be positive. The Y axis represents the percentage of the total observations that are selected in the three sets: all observations, positive observations from the training set, and positive observations from the validation set. To read values on the chart, hover over a location above the X axis.

☝ Hover over the X axis at the 20% location.

The green and red numbers above the hover point indicate the density of positive observations within the selected observations. In the ANN classifier, at the 20% level, 98.6% of the training observations (red) are positive and 96.3%

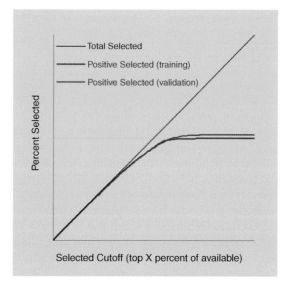

**Figure 5.12**   Lift Chart

of the validation observations (green) are positive. Thus, if we only select the top 20%, we would have a false positive rate of 1.4% on the training observations and 3.7% on the validation observations.

☞   Gradually move to the right along the X axis. Note the changes in selected positive training and validation percentages. What is the greatest "Selected Cutoff" percentage that you can achieve while maintaining a "100.0% positive" rate for the validation data?

The horizontal gray bar represents the total number of positive observations in the training set. As the red line approaches that bar, it becomes more and more difficult to select additional positive observations. Once all positive training observations have been selected, the red line flattens. The same is true for the green validation line.

As noted previously, the costs associated with false positives and the costs associated with false negatives in a classifier's application are usually different. If those costs can be associated with a monetary amount, then resulting monetary gains/losses can be plotted.

☞   Select the "$" button. A dialog box opens allowing you to specify monetary parameters that are applied to a simple cost/benefit model.

☞   For the Wheels example, enter the following:

| Field | Amount | Description |
|---|---|---|
| Fixed (setup) costs | $215,000 | 20,000 names × $10/name + $15,000 for brochure preparation |
| Cost per treatment | $222 | $12 for brochure + $210 for trial week |
| Net revenue per positive response | $900 | |
| Net revenue per negative response | | leave blank |
| Number (of potential treatments) | 20,000 | planned number of names to purchase |
| Expected positive rate | 8.2 | the percentage in the full dataset |

☞ Select "OK".

☞ Select "Profit Analysis" in the "Curve type" drop-down (Figure 5.13).

The **profit analysis** is similar to the lift chart. The X axis is the same although in this case it can be thought of as the top X% to be treated. The Y axis plots the expected net revenue gain based on the number of positive observations selected from the lift chart which are applied to the gain formula. at the 0% selected (treated) level, the loss is $215,000 – the fixed costs without any positive treatments gains to offset those costs. As the treatment percentage increases from zero, the profit begins to rise. At lower percentages, most of the selected

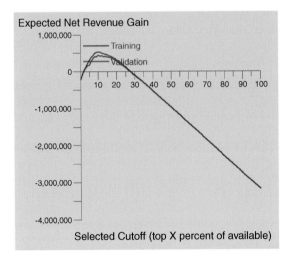

**Figure 5.13**   Profit Analysis

observations are positive and generate revenue. At some point, the proportion of positive treated observations relative to the total treated begins to fall. The $222 cost of treatment is applied to all, but there are fewer and fewer positives to bring in additional revenue. Eventually the expected net revenue gain drops below zero and continues downward as more and more non-buyers are offered the promotion.

In looking at the chart, the gain using the validation data appears to peak somewhere around 10% selected. You might ask why it would continue to increase after passing the 8.2% buyer level in the dataset. The answer is that when the top 8.2% of the observations are selected, some will be non-buyers due to misclassifications by the classifier. As the percentage treated is increased, there will still be some buyers selected. This will continue to boost the revenue gain until the increased cost of non-buyers wipes out any gains by the newly selected buyers. This appears to happen at about the 10% level.

☞ Compare the profit analysis based on your ann model with a profit analysis using the decision tree model. How do they compare?

## Classification Ensembles

Given the three classifiers modeling the same dataset, one might wonder if a combination of the three could collectively generate better predictions than any single model. In VisMiner, a combination of models is referred to as an **ensemble**.

☞ Right-click on the "Tree Classifier: Target" model.

☞ Select "Create classifier ensemble".

A new classification model is created with the name "Ensemble: Target". The single arrow from the tree classifier model to the ensemble model indicates that the tree classifier is the only participant in this newly created ensemble.

☞ Drag the "ANN Classifier: Target" model over the ensemble and drop.

With a second model added to the ensemble, VisMiner uses probability predictions from each model to generate a single combined prediction.

☞ Hover over the "Ensemble: Target" model to see its performance measures.

☞ Hover over the two contributing models to compare their performance measures with those of the ensemble.

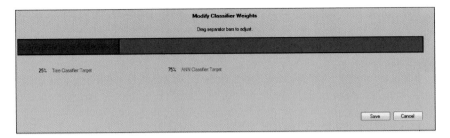

Modify Classifier Weights

Drag separator bars to adjust.

25% Tree Classifier:Target        75% ANN Classifier:Target

Save    Cancel

**Figure 5.14** Ensemble Weights

Although it won't always be the case, ensembles can produce results that outperform any of the contributing models.

By default, each model contributes equally to the class prediction. This can be overridden by the user.

☞ Right-click on the "Ensemble: Target" model.

☞ Select "Edit ensemble weights".

☞ Drag the dividing line between the red (tree classifier) and blue (ANN classifier) to the left until the tree classifier reaches the 25% weight level (Figure 5.14).

☞ Click "Save".

☞ Again hover over the "Ensemble: Target" model to review the performance measures.

Each time the weights of an ensemble are adjusted, VisMiner recomputes the performance measures. The weight distribution that maximizes a given performance measure is never a given. It will vary depending on the participating models and dataset. Obviously, the user can manually search for an optimal weight setting through a trial and error process. However, VisMiner does provide an automated search mechanism when there are exactly two contributing models to the ensemble. Access this feature by right-clicking on the ensemble, then selecting either "Find best (auc) weights" or "Find best (error) weights" to search the weight distribution that maximizes either the AUC or the error performance measure respectively.

## Model Application

If the only objective in performing data mining is to locate and understand the relationships between attributes, you will have completed that task using the features of VisMiner previously described in the tutorial. If, however, the objective is to apply the data mining knowledge discovered to future decision-making activities, then your best previously constructed models should be employed.

Suppose that Wheels has decided to decided to purchase additional potential customer data from Tell All and to use the ensemble model to choose individuals from the new dataset to receive their proposed promotional offer. How should it proceed?

The dataset PotentialBuyers.csv contains a list of 5,000 new individuals from Tell All.

☞ Open PotentialBuyers.csv.

☞ View the summary statistics.

The new dataset contains all attributes found in the original data except for the Buyer column. We will not know if individuals from this list are going to buy until we send them the offer and they have a chance to try out the car. Who should receive this offer?

☞ Drag PotentialBuyers.csv over the previously created ensemble model and release.

☞ Select "Generate predictions".

A new dataset named "Predicted:PotentialBuyers.csv" is created.

☞ To avoid such a cumbersome name, rename this dataset "PredBuyers".

☞ View the summary statistics for PredBuyers.

The newly generated dataset has all of the columns of the original plus three more: BuyerPredicted, NProb, and YProb.

BuyerPredicted is the class (Y or N) predicted by the model for each observation. Notice that there are 4,000 predicted as "N" and 1,000 as "Y". NProb and YProb are the likelihoods that the observation is "N" and "Y" respectively.

☞ Use either the parallel plot or the Control Center's filtered dataset option to create a subset named "LikelyBuyers" of individuals where YProb > 0.90.

☞ Save the LikelyBuyers dataset.

☞ View the summary statistics.

How many rows are in LikelyBuyers?
What percentage of the original 5,000 individuals are in this subset?
Congratulations, you now have a dataset containing only very likely car buyers.

# Summary

Classification models are used to predict a categorical (class) output value based on a given set of inputs. VisMiner implements three methodologies for building classification models: decision trees, artificial neural networks, and support vector machines. Each methodology has its strengths and weaknesses. Decision trees are easy to interpret and understand, but frequently do not provide the power of discrimination of the other two. Support vector machines discriminate extremely well, but frequently overfit the training data – producing models that do not generalize well. The effectiveness of each depends on the data. It is prudent to build all three types of models, then choose the model that performs best.

There are five steps to building a classification model:

1. Determine the prediction or output attribute.

2. Select input attributes based on their potential to contribute to model performance.

3. Prepare a dataset containing the input and output attributes. Split the dataset into training and validation observations. You may also want to change the ratio of output values in the dataset depending on their original balance.

4. Apply a modeler to the dataset to build the model.

5. Evaluate the resulting model.

Model evaluation should focus on the predictive ability of the model. It should also assess the contributions of input attributes to output probabilities.

A simple measure of predictive ability is error rate. Evaluations should consider the overall error rate as well as the rates of individual error types – false positives and false negatives in binary classifiers. The costs of each type of error may be quite different depending on the model application.

In addition to the error rate, evaluate the certainties with which models make their predictions. In VisMiner, receiver operating characteristic curves, lift chart, and profit analyses support this analysis.

To gain model understanding, look at the contributions that attributes make both individually and in combination to the classification process. In VisMiner, the decision tree graph enhances this understanding as well as the probability surface plots of the ANN and SVM models.

## Exercise 5.2

Using all CarBuyer observations (do not balance the Y and N frequencies), build ANN, decision tree, and SVM models. Use the same selected attributes as the previous tutorial (Age, Buyer, CreditRating, Gender, HomeLocation, Income, JobCategory, MaritalStatus, NbrSpeedingCitations, NbrSportsCars, Region). In building the derived dataset, specify 15,000 training observations and 5,000 validation.

1. How do the error rates of each model compare between these latest models and between these models and the earlier models built using the balanced dataset?

2. Conduct a profit analysis using the same parameters. How do the models compare between each other and between the previous models? If the promotion was to be actually implemented, which model would you use? Why?

## Exercise 5.3

The Mushroom dataset contains characteristics and measures of just over 8,000 mushroom samples. The dataset also contains an Edibility attribute that is either "edible" or "poisonous". Use Edibility as your output attribute.

1. Using the available visualizations, select attributes to be used for input to your classifier. Try to limit your selection to at most five input attributes. Create a derived dataset containing just these attributes. Which attributes did you choose? Explain why each was chosen.

2. Build classification models using the ANN, decision tree, and SVM modelers. How do the error rates compare? Rank them in order from best to worst.

3. There are two types of errors: an edible mushroom that is classified as poisonous and a poisonous mushroom that is classified as edible. If the classifier is to be used to determine which mushrooms to include in your lunch salad, which type of error is worse? Why?

4. View the ROC curves and lift charts of each of the three models. Select Poisonous as the positive result. Rank in order from best to worst according

to "area under curve". Is the ranking based on AUC different from the ranking based on error rate?

5.  Remember, when using a model, you do not need to go with the model's predicted value. Instead, you may specify your own probability cut-off upon which you will base your action. In the case of the mushroom classifier, how might you use the model to virtually eliminate the possibility of consuming a poisonous mushroom?

# 6

# Regression Analysis

**Regression analysis** is a form of prediction modeling that uses selected input attribute values to predict an output value. The difference between classification (covered in chapter 5) and regression analysis is in the output data type. In classification, the output or predicted type is nominal; in regression the predicted type is continuous numeric.

## The Regression Model

A formal definition of the regression model is:

$$Y_i = f(X_{1i}, X_{2i}, \cdots X_{pi}) + \varepsilon_i$$

where:

$Y_i$ is the value of the output variable for the $i^{th}$ observation
$X_{1i}$ is the value of the first input variable for the $i^{th}$ observation
$X_{pi}$ is the value of the $p^{th}$ input variable for the $i^{th}$ observation
$\varepsilon_i$ is the random error term for the $i^{th}$ observation.

The error term is included for two reasons.

1. In most data collection processes there are imprecisions and errors in measurement.
2. The model as defined is not complete. It is an abstraction of the real world. There are very likely to be additional inputs that influence the output yet are not included in the model and the form of the function itself may not accurately represent real world processes.

*Visual Data Mining: The VisMiner Approach*, First Edition. Russell K. Anderson.
© 2013 John Wiley & Sons, Ltd. Published 2013 by John Wiley & Sons, Ltd.

As with classification, to construct a regression model, a dataset containing data from past events is used. That dataset needs to contain values for both the input and output attributes. Once constructed and validated, the model can be used in the future to make predictions when the input attribute values are known, but the value of the output attribute does not yet exist or is unknown. For example, a manufacturer may want to predict sales based on planned advertising, pricing, and other inputs; an insurance underwriter may want to predict future expected loss amounts; a bookie may want to predict the point spread of a sporting event; or an economist may attempt to predict economic growth.

As in any data mining application, another potential use of regression analysis is simply to gain a better understanding of relationships between input attributes and the targeted output attribute. For example, a marketing research analyst may wish to better understand, and potentially quantify, the relationship between advertising and sales. What is the contribution of each new advertising dollar to sales; and is the contribution constant or are there decreasing returns beyond a certain level? Is there a threshold under which the level of advertising does very little or no good? A home building company, in evaluating potential new home designs, may want to know if the addition of a formal dining room would increase the value enough to cover the added cost.

A more difficult assessment when evaluating relationships is in the identification of interactions between inputs. For example, the response to advertising may be different between potential female and male customers; or the benefits of a dining room are probably different with respect to a 1000 square foot house versus an 8000 square foot house.

## Correlation and Causation

As a reminder before proceeding further, don't forget that just because a significant relationship is found between a pair of input and output attributes, it does not mean that the input attribute causes the output. It may be that the opposite is true – a change in the output variable causes a corresponding change in the input variable; hence the relationship exists. For example, if a positive relationship is observed between diet Coke consumption and obesity, does that mean diet Coke consumption leads to obesity? Or might it be that being obese leads one to consume more diet Coke?

In other situations it may be that a third unmeasured attribute causes both. For example, when a relationship exists between shoe size and basketball playing ability, does that imply that wearing bigger shoes will improve one's playing ability or that increased practice on the basketball court will force you into bigger shoes? The answer to both is obviously "no".

The only theoretically sound source from which definitive statements about cause and effect is data collected in a well-designed and tightly controlled experimental setting. When working with field data, the source of almost all data mining content, statements of causation may only be surmised based on hypothetical models of the processes under study.

Given that data mining, by definition, begins without any preconceived hypotheses, be wary of conclusions derived from the patterns uncovered.

## Algorithms for Regression Analysis

A simple, well-studied, popular, and long-used algorithm for regression is linear regression. In its simplest form, the model is defined as follows:

$$Y_i = \beta_0 + \beta_1 X_i + \varepsilon_i$$

where:

$Y_i$ is the value of the output variable for the $i^{th}$ observation
$X_i$ is the value of the input variable for the $i^{th}$ observation
$\beta_0$ is the Y intercept
$\beta_1$ is the slope or coefficient of input variable $X$
$\varepsilon_i$ is the random error term for the $i^{th}$ observation.

This model is usually referred to as **simple linear regression**, because it has only one input. Linear regression can be extended with more input variables as:

$$Y_i = \beta_0 + \beta_1 X_{1i} + \beta_2 X_{2i} + \ldots + \beta_{p-1} X_{p-1i} + \varepsilon_i$$

where:

$p - 1$ is the number of input variables. Note: $p - 1$ is typically used to refer to the number of input variables as there is one more implied input $X_0$ whose value is always 1. Hence, a total of $p$ variables when $X_0$ is included.

With multiple inputs, the model is referred to as **multiple linear regression**. As the model has been defined, the task of multiple linear regression becomes that of generating estimates for all $\beta$ coefficients using a dataset containing values for both the output variable and all input variables. VisMiner employs a common method for generating these estimates: **ordinary least squares**. A description of the method is not included here. For the interested reader, there are numerous books and websites available describing the method.

## Assessing Regression Model Performance

Regression model performance is defined as a measure of how well the model predicts the output for a given input. In other words, how small is the error term?

Since "smallness" of error is relative, a benchmark or base model is deployed, against which our model is compared. In regression, that base model is the mean of the output variable $Y$. That is, if we don't use any other inputs to estimate $Y_i$ for the $i^{th}$ observation then simply use the mean. When an existing dataset with known $Y$ values is applied to this base model, we calculate the total error of the model as:

$$SSE = \sum_{i=1}^{n} (Y_i - \bar{Y})^2$$

where:

SSE is the sum of squares – the measure of total error of the prediction model
$n$ is the number of observations in the dataset
$Y_i$ is the actual output value of the $i^{th}$ observation
$\bar{Y}$ is the mean output value.

When the same dataset is applied to a regression model, its sum of squares is:

$$SSE_{reg} = \sum_{i=1}^{n} (Y_i - \hat{Y}_i)^2$$

where:

$\hat{Y}_i$ is the output value predicted by the model for the $i^{th}$ observation.

If the error of the regression model ($SSE_{reg}$) is less than the error of the base model ($SSE$), then it has improved upon the base model. A measure quantifying the improvement is $R^2$:

$$R^2 = 1 - \frac{SSE_{reg}}{SSE}$$

It is usually referred to as the **coefficient of determination**. In simple terms, $R^2$ is the percentage of variance in the data explained by the regression model. For example, if $SSE$ is 100 and $SSE_{reg}$ is 10, then $R^2$ is 0.90 or a 90% improvement in prediction capability over the base model. A regression model that perfectly predicts the output value will have a regression error of zero and $R^2$ of 1.0 (100% accuracy).

Although the ideal $R^2$ is 1.0, one may wonder what a minimum acceptable $R^2$ might be. Any significant (not likely to have happened by chance) value of $R^2$ above 0 is an improvement over nothing. However, when comparing regression models, built using the same input attributes and values, the model with the higher $R^2$ would always be preferred.

When comparing models built using the same input observations but where the input attributes of one model are a subset of the input attributes of the other, an **adjusted $R^2$** is frequently computed. The adjustment to $R^2$ will factor a penalty into the model built using the greater number of input attributes. Its purpose is to overcome the tendency for $R^2$ to increase due to chance with the addition of new attributes even when those attributes do not make any contribution with respect to the total population.

## Model Validity

The $R^2$ statistic of a regression model is an indicator of how well the model performs. That is, how well do the input values explain the variations in the output values? In regression analysis, a related question is, how valid are the results? Do the non-zero valued coefficients generated actually represent relationships between the input and output columns, or is it possible that the coefficients are chance occurrences resulting from the incompleteness of the dataset?

## Looking Beyond $R^2$

Consider the four datasets in Figure 6.1, which were created by Anscombe in 1973 to illustrate possible pitfalls in linear regression. [Anscombe, F. J. "Graphs in Statistical Analysis". *American Statistician* **27** (1): 17–21]

When a simple linear regression modeler is applied to each of the four datasets using *y* as the output column, the resulting models are nearly identical.

| x | y |     | x | y |     | x | y |     | x | y |
|---|---|-----|---|---|-----|---|---|-----|---|---|
| 10 | 8.04 | | 10 | 9.14 | | 10 | 7.46 | | 8 | 6.58 |
| 8 | 6.95 | | 8 | 8.14 | | 8 | 6.77 | | 8 | 5.76 |
| 13 | 7.58 | | 13 | 8.74 | | 13 | 12.74 | | 8 | 7.71 |
| 9 | 8.81 | | 9 | 8.77 | | 9 | 7.11 | | 8 | 8.84 |
| 11 | 8.33 | | 11 | 9.26 | | 11 | 7.81 | | 8 | 8.47 |
| 14 | 9.96 | | 14 | 8.10 | | 14 | 8.84 | | 8 | 7.04 |
| 6 | 7.24 | | 6 | 6.13 | | 6 | 6.08 | | 8 | 5.25 |
| 4 | 4.26 | | 4 | 3.10 | | 4 | 5.39 | | 19 | 12.5 |
| 12 | 10.84 | | 12 | 9.13 | | 12 | 8.15 | | 8 | 5.56 |
| 7 | 4.82 | | 7 | 7.26 | | 7 | 6.42 | | 8 | 7.91 |
| 5 | 5.68 | | 5 | 4.74 | | 5 | 5.73 | | 8 | 6.89 |
| (a) | | | (b) | | | (c) | | | (d) | |

**Figure 6.1** Regression Datasets

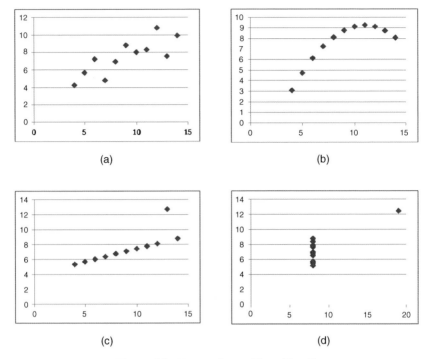

(a)                                    (b)

(c)                                    (d)

**Figure 6.2**    Dataset Scatter Plots ($X$ vs $Y$)

The value of $R^2$ is 0.67, the intercept is 3.00 and the coefficient (slope) of $x$ is 0.50. Yet, when one compares scatter plots of the four datasets, each set has its own unique characteristics (see Figure 6.2).

Dataset (a) appears to contain points scattered around an upward sloping imaginary line. This is what one would typically expect to find in a dataset with a linear relationship between variables. In dataset (b), the points appear to perfectly fit a curve. The perfect fit indicates that a linear model is not the best to fit the data and that $R^2$ may be improved by fitting a curvilinear model. Dataset (c) looks like it should generate a perfect linear fit except for the one outlying point. If the outlier is removed, $R^2$ should be 1.0. Dataset (d) does not appear to have a distribution that would support a fit to any kind of model. All but one of the $x$ values are 8. It is this single outlier that dictates the slope and intercept of the fitted line. If, for example, it had a $y$ value of 2.5 instead of 12.5, the slope would be negative instead of positive as the fitted line would pass through this point no matter where it appeared on the Y axis.

The examples of Anscombe highlight the need to explore and understand the nature of the data before choosing and applying a modeling technique. Dataset (b) needs to have a non-linear modeler applied; in dataset (c), outliers should be removed before processing; and since dataset (d) does not suggest any kind of relationship, no modeling at all is recommended. The dataset viewers of VisMiner are a good tool to begin this initial exploration and analysis.

## Polynomial Regression

To convert a linear regression modeler into a non-linear modeler, a common method is to add additional columns that are non-linear transformations of the original input columns. In a second-order polynomial regression, the inputs are squared. For example, the simple linear regression model becomes:

$$Y_i = \beta_0 + \beta_1 X_i + \beta_2 X_i^2 + \varepsilon_i$$

A second-order polynomial regression using two or more numeric input columns would add squared terms for each.

A third-order polynomial regression becomes:

$$Y_i = \beta_0 + \beta_1 X_i + \beta_2 X_i^2 + \beta_3 X_i^3 + \varepsilon_i$$

Again, when multiple numeric input columns are included in the model, each input would be expanded to include both the squared and cubed terms. A similar expansion is possible for polynomial regressions of even higher order.

## Artificial Neural Networks for Regression Analysis

To model more complex relationships beyond that provided by polynomial regression, artificial neural networks (ANN) are a good choice. Using ANN processors, models can be constructed to fit almost any relationship. They are especially good at modeling relationships where input attributes interact. Input interaction occurs when the levels of one input attribute change the relationship of a second input attribute with respect to the output attribute. For example, in a particular setting, it may be that the effect of advertising on sales is different for men and women. In this case, the input attribute "Gender" changes the pattern of the contribution of input attribute "Advertising" on output attribute "Sales".

## Dataset Preparation

The dataset used for regression modeling in VisMiner must be in a tabular format. The input attributes may be of any data type – numeric, ordinal, or nominal. The output attribute must be numeric.

It is important to remember that when using VisMiner to build a regression model, the dataset should contain only the attributes (input and output) to be used by the modeling process. There should not be any row identifiers or other attributes included. Therefore, if your dataset contains unneeded attributes, before starting the modeler, create a derived set containing only the attributes that you want to include. See Chapters 2 and 3 for details on how this is done.

## Tutorial

VisMiner supports the two main objectives of regression analysis in data mining:

1. Construct a model that accurately predicts continuous numeric output values using input datasets that were not used to build the model.
2. Learn more about the relationships between potential predictors and the target output attribute.

With respect to the first objective, in regression analysis, the best overall measure of model performance is $R^2$. The recommended method for comparing alternative models and assessing their ability to generalize is to compare the $R^2$ values of the models when applied to test or validation datasets deemed to be representative of the datasets to which the model will be used.

The task of building that best model is one of choosing the right combination of input attributes, then applying those input attributes to a modeling algorithm that will best meet the objectives – prediction accuracy and model understandability.

VisMiner implements three algorithms for regression analysis: linear regression, polynomial regression, and artificial neural networks (ANN).

From the perspective of interpretability, linear regression is the simplest of the three, followed by polynomial regression and ANN regression. ANN based models are by far the most difficult to evaluate, to understand, and to visualize their inner workings.

On the other hand, ANNs are the most powerful with respect to their ability to fit the data and thus generate accurate predictions. The ANN algorithm is the only one of the three that can detect **interactions** between input attributes. ANNs and polynomial regression modelers are the two that can do non-linear fits of the data. In order to fit non-linear inputs to linear regression models, user defined and initiated transformations of attributes are required prior to submission of the dataset to the modeler.

In conducting a regression analysis using VisMiner, it is recommended that all three algorithms be applied. Although, in the end, the ANN will likely generate the best model, the linear and polynomial regression modelers can assist in choosing predictor attributes, gaining insights into their contributions to the predicted output, and providing benchmarks against which the performance of competing models can be judged.

# A Regression Model for Home Appraisal

In Chapter 3, we visually explored the dataset of homes for sale in the Provo, Utah, metropolitan area. After handling missing values, the data was saved in the file CmpltHomes.csv. We will use this dataset to evaluate the influence of selected inputs on home sale price and to generate appraisals of home values. Given that sales price is a continuous numeric value, we use regression analysis tools to perform this evaluation. (Note: It is recognized that the dataset deployed herein contains asking prices not actual sale prices. However, for lack of a better dataset, we will treat them as if they were the actual sale prices.)

The first step in regression analysis is to select the input attributes to include in the model. Although there are algorithms to do so, VisMiner does not provide features supporting the automatic selection of input attributes. In designing VisMiner, it was felt that the manual process of reviewing and selecting attributes for inclusion was extremely valuable and would contribute to better appreciation of the system being modeled and the rationale for the inputs used in the final model.

We can start by eliminating all attributes which, from a common-sense perspective or from a suitability perspective, should not be used.

With respect to suitability, state should be eliminated because all observations have the same value (UT). The attributes city, elementary, jrHigh, and street are nominal data types with high cardinalities. When using nominal data in regression modeling, they must first be converted to a set of dummy variables. VisMiner does the conversion automatically, but limits acceptable nominal attributes to those with a cardinality of 10 or less. The attributes just listed all exceed that limit.

From a common-sense approach grounded in domain knowledge, one should question the use of daysOnMarket as a predictor. It could be argued that daysOnMarket is dependent on price rather than the other way around. Also, the attribute zip is stored in the dataset as an integer and would thus be used by a modeler as numeric. Yet zip codes, even though stored as a number, are actually nominal data types. To be used properly it would need to be converted, after which its cardinality of 23 would be a problem. It could also be argued that the potential contribution of zip is covered by other location-based attributes: latitude, longitude, and schoolDistrict.

# Modeling with the Right Set of Observations

Look back at Figure 6.2c. Outliers in the dataset will bias the results.

We have previously defined and described methodologies for identifying and eliminating outliers (Chapter 3). However, even though observations may not technically be classified as outliers, there may be observations in a dataset that lie outside the range of interest of a particular data mining exercise. For example, a biologist studying the sleep habits of small mammals would not want to include any measures belonging to large mammals in the dataset even though the data is valid.

↪ Open the CmpltHomes.csv dataset.

↪ View the dataset in a parallel coordinate plot.

↪ For readability, you may want to drag all axes belonging to nominal attributes to the right to keep their labels from obscuring the numeric attributes of interest.

Look at the distributions of lot and price. Almost all homes in the dataset have lots less than one acre in size, while there are a few homes with lots ranging up to 200 acres. Most of the homes are priced under about $750,000. The highest priced home is $8.9 million. Even though the data of these large lot or high priced homes may be valid, they may be outside our area of interest with respect to our data mining objectives. Like the single observation in Figure 6.2c, these observations, if included in the analyses, may bias the models we generate.

Suppose that the reason for generating estimates of a home's sale price is to identify homes for purchase as an investment. If the predicted price is well above the asking price, then the potential for a good return on the investment is greater. Suppose also that the investors have a policy to only invest in homes on lots under two acres, priced less than $750,000, and under 5,000 square feet in size. Including observations in the model building process outside these restrictions risks the construction of models that are biased by these observations.

↪ Use the parallel coordinate plot or the Control Center's "Create filtered dataset" option to eliminate all homes with lots over two acres, priced over $750,000, or over 5,000 square feet in size.

↪ Name this set "selectedHomes".

↪ Create a dataset derived from selectedHomes named "homes" containing only those attributes deemed acceptable to the regression modeling process. (Include all but: city, daysOnMarket, elementary, jrHigh, state, street, and zip.)

Note: in the tutorial that follows, measures and statistics reported may differ slightly from those that you experience as you follow along. This is because when filtering out observations, your slider positions may have been slightly different. The dataset (homes) in the examples below contained 3,111 observations. Yours may have a few more or a few less; however, your modeling results will not be significantly different.

The process of input selection can be approached from a **bottom-up** or a **top-down** perspective. In the bottom-up approach, attributes are evaluated and selected

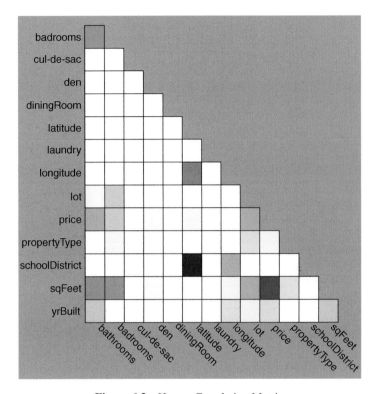

**Figure 6.3**   Homes Correlation Matrix

one at a time for inclusion in the model, based on their potential contribution. In the top-down approach, all acceptable attributes are at first included, then iteratively excluded as their contribution is found to be insufficient.

The correlation matrix is a good starting point for the bottom-up approach.

↳ View the homes dataset in a correlation matrix (Figure 6.3).

The two most highly correlated attributes with respect to price are sqFeet (0.713) and bathrooms (0.476). Begin the analysis using these two attributes.

↳ Create a derived dataset named "sqFtBath" containing just the attributes: bathrooms, price, and sqFeet.

The dataset is now ready for application of a regression algorithm.

↳ Drag the "Linear Regression" modeler and drop over the sqFtBath dataset.

↳ Choose price as the prediction column.

|          | Rows | R² | Adj R² |
|----------|------|------|--------|
| Training | 3111 | 0.519 | 0.519 |
| Validation | – | – | – |

**The Regression Model**

| Output Column | F Statistic |
|---------------|-------------|
| price | 1,679 |

| Input Column | Coefficient | P-Value |
|--------------|-------------|---------|
| Intercept | 34.957 | 0.0000 |
| bathrooms | 12.339 | 0.0000 |
| sqFeet | 53.6 | 0.0000 |

**Figure 6.4**   Regression Summary

VisMiner provides two viewers for evaluation of regression models. The first is the regression summary which gives a text-based summary of the model performance and validity.

☞   Drag the model LinearRegression:sqFtBath up to a display, selecting "Regression Summary".

The summary (Figure 6.4) is broken into three small tables. The first table provides the regression goodness-of-fit measure $R^2$ and the $R^2$ value adjusted for the number of input attributes. For this particular model, the size of the house (sqFeet) and the number of bathrooms explain about 52% of the total variation in price. Note also that all rows were used to build the model. At this point, none were set aside for model validation. We'll do that later.

The second table contains the F statistic for the regression (1,679). Given two degrees of freedom for the numerator and 3,508 for the denominator, the likelihood that these inputs in actuality contribute nothing to price prediction is extremely low.

The third table provides a summary of the regression coefficients from which the model is constructed. The model is:

$$price = 34957 + 12339 \cdot bathrooms + 53.6 \cdot sqFeet$$

Each bathroom in a house contributes $12,339 to its value, and size contributes $53.60 per square foot.

The p-values for each coefficient are essentially zero; hence, we are quite sure that both square feet and number of bathrooms do actually contribute to price. Given the $R^2$ value of 0.519, it is also evident that the model is incomplete. Other attributes should be found that will contribute to the model's fitness.

To see the model visualized:

Drag the model LinearRegression:sqFtBath up to a display, selecting "Regression Model".

The regression model viewer plots the model on a 3-D XYZ graph. The output attribute is always represented on the Y axis. Input attributes of the model are represented on the X and Z axes. The viewer plots the model function. In the case of linear regression, that function is depicted as a plane. Note: A linear regression with more than two input attributes would define a hyperplane. However, the regression model viewer plots with respect to two input attributes only, thus always seen as a plane.

Looking at the plot (Figure 6.5) you see the steeper slope with respect to sqFeet than bathrooms. Relatively speaking across the range of values along each axis, home size contributes more to price than the number of bathrooms.

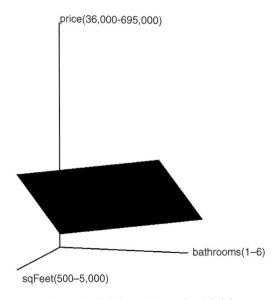

**Figure 6.5**  Linear Regression Model

In any regression analysis, making predictions with the model using input values outside the range of those used to build the model are not recommended. For example, the model as drawn shows a projected price for a six bathroom home with 500 square feet (the corner of the plot at the high end of the X axis). Yet, does such a home really exist? Probably not.

To see on the plot the range of home observations that was used to build the model.

☞   Check the "Surface Density" box.

When checked, the viewer uses opacity to indicate input observation density. Areas on the plane that did not have corresponding input observations during model construction are depicted as almost totally transparent. Evaluating the plot, you see that most of the observations lie near the diagonal region running from the origin out to the opposite corner of the plane.

☞   Check the "Plot Points" box to show the points used in building the model.

Most of the points lie close to the surface plane, which they should. There are some high-priced outlying homes lying well above the model surface. It is these homes that are contributing most to the total overall error of the model. You might wonder if a curvilinear model might fit better.

☞   Drag the "Polynomial Regr 2$^{nd}$ Order" modeler over the sqFtBath dataset and drop.

☞   Again choose price as the prediction column.

☞   View the regression summary for this model (Figure 6.6).

The squared term additions of the polynomial model increase $R^2$ from 0.519 up to 0.542, not a huge improvement. However, the coefficients have changed significantly. For example, the Y intercept went from 34,957 to 103,113.

☞   Compare the two models (linear and polynomial) in side-by-side regression model viewers.

In the polynomial regression, along the Z (sqFeet) axis, you see the curvilinear contribution reflecting the (sqFeet)$^2$ coefficient. Although, given the incompleteness of the model, it would be a careless generalization; one could say that the contribution of each additional square foot increases exponentially. On the other hand, along the X axis you see a nearly linear contribution of bathrooms with a corresponding (bathrooms)$^2$ p-value of 0.4246.

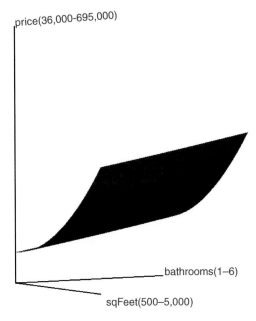

price(36,000-695,000)

bathrooms(1–6)

sqFeet(500–5,000)

**Figure 6.6**    Polynomial Regression

Exercise 6.1

Apply the third-order polynomial regression modeler to the sqFtBath dataset. Does it improve upon the second-order polynomial model? Explain your answer.

## ANN Modeling

The third regression algorithm available in VisMiner is the artificial neural network (ANN). The advantage of the ANN is that it is capable of fitting a more complex surface.

⟴  Drag the ANN regression modeler over the sqFtBath dataset and drop.

⟴  Choose "Build interactively", then select price as the prediction column.

The interactive build option allows user control over the training process. The effectiveness of ANN training depends on the learning rate and momentum, yet there is no single best learning rate or momentum for all datasets. It will vary from dataset to dataset. The interactive build option allows you to monitor the training progress while adjusting the learning rate and momentum as it trains.

The interactive build begins after the automatic completion of a single epoch pass using the learning rate and momentum represented by the red dot in the grid – a very high learning rate paired with a low momentum. The initial "Training R Squared" is shown above the "Training Progress Plot". Normally the initial $R^2$ will be negative meaning that its predictions are worse than when just the mean output value is used. It will improve as training progresses.

Each time you press the mouse button down while over the grid, training resumes using the learning rate and momentum corresponding to the current mouse location. It continues until the button is released. While training, you may drag the red dot to other locations within the grid to change the learning rate and momentum on the fly. The progress plot is updated to show the current value of $R^2$. When the mouse button is released, a checkpoint is created. Checkpoints record the current state of the ANN and are depicted on the progress plot by small red circles. They allow you to go back to previous checkpoints to try training in a different direction. The interactive process allows you to visually search for the right combination of learning rate and momentum applicable to the dataset.

☞   While holding the mouse down, slowly drag the red dot down from the upper left corner towards the bottom center.

As you drag, you should see the training progress plot remain flat ($R^2 < 0$), then gradually start to climb. While training, it is a good idea to release then press the mouse button every few seconds to set checkpoints that you may want to return to. Note: There will always be an initial checkpoint available which is set immediately after completing the first training epoch.

As you train, keep in mind the $R^2$ values achieved by the linear and polynomial regressions (0.519 and 0.542 respectively). If you cannot push the ANN $R^2$ past these values, it is not a useful model.

Continue the training process, pushing the red dot towards all regions of the grid. Don't worry if the $R^2$ value drops below zero while experimenting, it will quickly rebound when moved to a more productive location. With this particular dataset, you will find that you can get $R^2$ up to almost 0.55.

☞   Click "Accept" when you have reached the target $R^2$ above 0.54.

☞   Compare the second-order polynomial and ANN regression models in side-by-side regression model viewers.

In the polynomial model (Figure 6.6) notice the consistently increasing slopes produced by the sqFeet squared term of the modeler compared to the somewhat irregular curves of the ANN model (Figure 6.7). This is one of the reasons that the ANN modeler potentially generates better $R^2$ scores. Notice also the shape of the

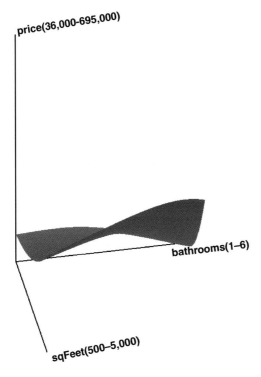

**Figure 6.7**   ANN Regression

curves on the opposite edges of the surfaces. For example, look at the curve along the X (bathrooms) axis where sqFeet = 500. It shows a negative contribution of each additional bathroom. In other words, for a small home, squeezing in more bathrooms reduces its value. Look at the curve along the opposite edge (sqFeet = 5,000). It shows a positive contribution for the first four or five bathrooms, then a negative return for the sixth. When the shapes of the curves at opposite edges of the surface differ, the modeler has detected interaction between the input attributes. We will see more dramatic differences in later models.

Given the relatively low levels of $R^2$ achieved by the modelers using sqFeet and bathrooms, the question should be asked as to which additional input attributes would improve the fitness of the model or if some other changes could be made.

Return to the correlation matrix to identify the top four attributes with respect to correlation with price.

☞ Create a derived dataset named "best4" consisting of price and the four most correlated attributes: sqFeet (0.713), bathrooms (0.476), lot (0.414), and bedrooms (0.365).

☞   Apply the linear regression modeler to best4.

☞   Apply the second-order polynomial regression modeler to best4.

☞   View the summary statistics for the two models (linear and polynomial regression).

After adding the two additional attributes to the model, $R^2$ for the linear model improved from 0.519 to 0.537 while for the polynomial model, it went from 0.542 to 0.563. The additions contributed to the construction of a better performing model, although the improvements were marginal.

The logical next step might be to add additional attributes, checking their contribution to model performance as we go. However, that can be left as an exercise. Instead we switch to a top-down approach.


## The Advantage of ANN Regression

Before continuing with the "homes" analysis, we are going to briefly switch datasets in order to more definitively illustrate the issues of interaction.

☞   Open the dataset xyRegress.csv.

☞   View the summary statistics.

There are three variables in the set: numeric X and Y columns, and the XNominal column with values "Y" and "N".

☞   Drag the linear regression modeler over the xyRegress dataset.

☞   Select Y as the prediction column. That is, use X and XNominal to predict Y.

☞   Hover over the model to view the results.

$R^2$ is less than 0.01. Obviously, a linear model does not fit the data.

☞   Build a third-order polynomial regression.

☞   Hover over the model to view the results.

$R^2$ barely improves.

☞   Interactively build an ANN regression model. Only about 50 epochs will be needed to train the model.

☞   Hover over the model to view the results.

With your ANN model, $R^2$ should have soared to about 0.98. Why the improvement?

⌨ View a scatter plot of xyRegress.csv.

⌨ Select X for the X axis, Y for the Y axis, and XNominal for the category.

This is a manufactured dataset to make a point about interaction effects. You should recognize in the scatter plot that the relationship between X and Y is very different for the XNominal = "Y" observations as compared to the XNominal = "N" observations. Additive modelers such as linear and polynomial regression cannot detect the interaction, whereas the ANN has no trouble doing so. This is its primary advantage. Obviously, you are very unlikely to see such a dramatic effect in a real world dataset. Yet it does exist in many datasets and your modeler needs to be able to detect that interaction.

## Top-Down Attribute Selection

The homes dataset contains all attributes previously determined to be candidates for inclusion in the regression models: price, bathrooms, bedrooms, cul-de-sac, den, diningRoom, latitude, laundry, longitude, lot, propertyType, schoolDistrict, sqFeet, and yrBuilt.

⌨ Build linear, second- and third-order polynomial regression models for this dataset.

⌨ View the regression summaries of each.

Using all available legitimate attributes, the linear model generates $R^2$ of 0.622 and the polynomial models generate 0.664 and 0.672 respectively. This gives us a target to try to exceed when building an ANN model.

⌨ Using the same dataset, interactively build an ANN regression model. You will find that it peaks out with an $R^2$ near 0.79.

Why is the performance of the ANN model better than the other two models? One answer is that an ANN potentially fits the data better and a primary reason for fitting the data is that the ANN can detect and model interaction between inputs whereas the linear and polynomial models do not. By interaction, we mean that the contributions of one input attribute vary depending on the value of a second input attribute. The linear and polynomial models are additive. The level of one input attribute in no way affects the contribution of a second input attribute. Yet in the real world, there are frequent interactions between inputs. For example, would you expect the value of a third bathroom in a one-bedroom home to be the

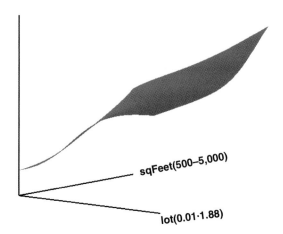

**Figure 6.8**   ANN Regression Interaction

same as that third bathroom in a six-bedroom home? In other words, does the value of a bathroom change depending on the number of bedrooms in a home? To detect interaction:

☞ View a regression model of the ANN model.

☞ To facilitate comparison, view a regression model of the third-order polynomial model to the side of the ANN model.

☞ On both plots, select sqFeet for the X axis and lot for the Z axis.

We previously presented the concept that interaction exists between input attributes when the shape of the curve along one edge of the surface differs from the shape along the opposite edge.

Looking first at the polynomial model, since the polynomial modeler cannot model interaction, the opposite edges of the model surface will have the same shape. The ANN model is shown in Figure 6.8. Compare the edge running from the origin out along the Z axis to the opposite edge. The edge running from the origin (sqFeet = 500) along the Z axis shows a lot size contribution that increases, then levels off. Along the opposite edge (sqFeet = 5,000) the contribution of lot size is a steady consistent increase. Clearly the contribution of lot size to price interacts with home size (sqFeet).

## Issues in Model Interpretation

In reviewing the ANN model plot, it is important to remember that we are seeing a plot of the model surface depicting how variations in lot and sqFeet affect price. All other inputs are held constant. The values of these other inputs are represented in the sliders to the right of the plot. When the plot is first viewed, the sliders are set to the mean value for numeric input attributes and to the modal value for nominal inputs. For example, in the plot of Figure 6.8, bathrooms is set at 2.36. Therefore, the model surface of the plot represents homes with 2.36 bathrooms only.

To see the contribution of bathroom count with respect to price, try slowly dragging the bathrooms slider from the low (left) end at 1.00 to the right. As you drag, the surface twists and changes shape. Some areas rise and some fall. This is another indication of interaction between bathrooms, lot size, and square feet.

One way to see the contribution of nominal attributes is to toggle back and forth between values while noting changes in the plot surface. For example, toggling den between "N" and "Y" helps to understand the contribution of a den to a home's price.

Another way to see the contribution of nominal attributes is to select that attribute in the "Category" drop-down.

⌘ Select propertyType in the category drop-down.

The resulting plot is in Figure 6.9. It shows that for almost all combinations, manufactured homes contribute the least to a home's value, while single family homes generally contribute the most. Remind yourself again that this is holding all other inputs constant.

The reader should hereby be warned that any attempts to assess the contributions of input attributes using two- or three-dimensional plots representing one or

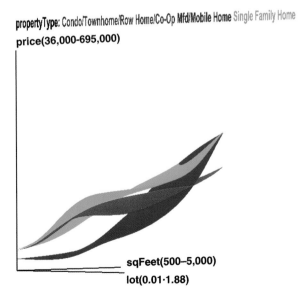

**Figure 6.9** Regression Surface by Property Type

two input attributes while holding all other inputs constant, is a very complex task when the model under review contains interactions between inputs. The model itself and the interactions it captures may be valid, yet a full review of all possible combinations of input levels is impossible. In the example dataset, there are millions of possible combinations. The linear and polynomial models, where all inputs contribute independently, are far easier to interpret.

One possible approach to getting around the complexity resulting from numerous input attributes in the ANN model is to build models containing only a small subset of attributes (for example, three or four inputs). Explore the limited interactions between these attributes. The $R^2$ performance will be lower, yet these smaller models may provide sufficient insights into the interactions to assist in gaining a better grasp of the environment under study as a whole.

In summary we could say: when model performance (prediction accuracy) is important, build the larger, more complex models; when the objective is understanding, build and interpret several smaller models.

## Model Validation

The last completed ANN regression model achieved a respectable $R^2$ slightly above 0.78. However, one must wonder if the model is not overfit with respect to the training data. If the model is overfit, it will not generalize well. That is, it will not generate accurate predictions when applied to future datasets.

To assess the generalizability of a model and as a guide to the model building process, validation datasets are employed. Typically a validation dataset is a portion of the original dataset that is held back from the training algorithm. It may be applied to the model after construction is complete to provide an indicator of how the model will perform using future datasets or it may be applied at evaluation points during construction to determine when the model is beginning to overfit. Once that overfit state is recognized, the training should be stopped.

⮕ Right-click on the homes dataset; select "Create derived dataset".

⮕ Give it a name of "trnValHomes". It will be used for both training and validation.

⮕ "Select All" columns.

⮕ Enter 1300 as the "Rows for validation set".

⮕ Click "Create".

Even though it shows in the Control Center as one dataset, we have now split the homes dataset into two partitions – 1,300 observations (about 40%) to be used for validation and the rest for training. The rows were randomly assigned to each partitions.

⏷ Open an interactive ANN regression modeler for the trnValHomes dataset selecting price as the prediction column.

As before, the ANN modeler uses just the training data to complete one epoch. At the end of this epoch, $R^2$ with respect to the training data is computed and then again with respect to the validation data. Both are displayed above the training progress plot. At this point after one epoch, the model is very unlikely to be trained. Given the initial random neuron weights, $R^2$ is likely to be negative.

⏷ Click "Decrease" to reduce the epochs per step to 1.

⏷ Click "Slower" about seven or eight times.

The purpose of the above two steps is to slow the training process down, allowing you, the user, to react in time to changing states in the model. If after training a few epochs, you feel able to respond fast enough, you may want to adjust the training speed upward.

The overall objective of the process is to find a training state with the best possible validation $R^2$. The challenge of finding this state is mostly a trial and error process. The best location and direction of movement on the training grid is dependent on the attributes and values of the dataset. It will change from dataset to dataset. The best strategy is to try moving in different directions on the grid. Whenever, you get to a point where the training $R^2$ is steadily increasing while the validation $R^2$ is decreasing, the ANN is overtrained. Go back to a previous location and try a different direction. Keep doing this until you feel that you have found a state that is difficult to improve upon. In testing that we have conducted using this dataset, we have been able to push the validation $R^2$ up to about 0.52 while at the same time $R^2$ for the training set has reached 0.79. Of course, $R^2$ for the training set could be pushed even higher at the expense of the validation $R^2$.

## Model Application

If our objective in building the model is to locate homes for investment purposes that have a good potential for appreciation, our approach is to use the model to generate predicted prices for all homes in the initial dataset. Once they are generated, we can compare actual asking price to predicted price, selecting the best bargains for additional investigation. Obviously there are characteristics and conditions of homes not captured by the dataset that would require a physical inspection to assess. However, using the model is a good starting point in identification of the best candidates for inspection.

⏷ Keep your newly created, best performing ANN regression model open.

⏷ Open the previously created SelectedHomes.csv dataset.

⊂̥  Drag SelectedHomes down over the ANN regression model and release.

⊂̥  Select "Generate predictions".

When a dataset is dropped on a model, if VisMiner finds all model input attributes in the dropped dataset, it will use those inputs to generate predicted output values for all rows, then create a new dataset containing the original data plus a new attribute having the same name as the output attribute followed by the word "Predicted". In this case the new name is "pricePredicted".

⊂̥  Change the name of this new dataset to an easier "predictedPrice".

⊂̥  Right-click on the dataset; select "Create derived dataset".

⊂̥  "Select All" existing attributes.

⊂̥  Create a new column named pcntGain with the formula: (pricePredicted − price)/price∗100.

⊂̥  Name the dataset "gains".

⊂̥  Click "Create".

⊂̥  Open gains in a parallel coordinate plot.

⊂̥  Drag the bottom slider of the priceGain column up until only 50 observations are visible (or as close to 50 as the display precision will allow you to get).

⊂̥  Save this dataset as "top50".

You now have a dataset containing the 50 best bargains according to the model built to predict price.

# Summary

Regression analysis is used to build models that generate predictions for a continuous numeric output attribute.

VisMiner implements three regression algorithms: linear regression, polynomial regression, and artificial neural networks (ANN). Of the three, the ANN algorithm is the only one that can detect interaction between input attributes.

In comparing alternative regression models, a common measure of model performance is $R^2$, which reflects the percentage of variation in output values explained by the model. To assess the contributions that input attributes make toward a predicted output value, VisMiner implements regression model surface plots. The surface plots are also very useful in detecting and evaluating interaction effects between inputs.

# 7

# Cluster Analysis

## Introduction

**Cluster analysis** is the process of grouping observations based on similarity (visually observed as proximity), connectedness, or density. The results of a cluster analysis are called a **clustering**.

Cluster analysis is similar in concept to the previously discussed process of classification. In classification, the observation groupings (classifications) are known *a priori*. The objective of classification analysis is to discover relationships between other dataset attributes and the previously known class attribute that could be used to predict class membership. However, in cluster analysis the groupings are not previously known. The objective is the discovery of clusters of observations grouped according to dataset attribute values.

In data mining, there are a number of potential objectives in conducting a cluster analysis.

- Sub-population identification and isolation. As has been discussed in previous chapters, datasets may be composed of observations drawn from populations with different characteristics. Relationships found only in a single subset may not be as readily identified when exploring the full dataset versus just the subset. Hence, a good rule of thumb is to isolate the subsets and then analyze individually. A strategy in product marketing is to first segment the market, then develop specific promotions for selected market segments. The same principle may be applied to data mining – isolate subsets, then develop custom analysis plans for each.

*Visual Data Mining: The VisMiner Approach*, First Edition. Russell K. Anderson.
© 2013 John Wiley & Sons, Ltd. Published 2013 by John Wiley & Sons, Ltd.

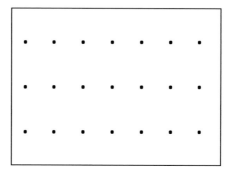

**Figure 7.1**   Connectedness

- Understanding. Evaluating differences between clusters may provide insights into the entire population and can guide future data analysis directions. Biologists have learned, for example, that it is not advantageous to study all forms of life as a whole, rather it is better to create taxonomies, then study individual groupings within the taxonomy.

- Data aggregation. When mining very large datasets, observations within clusters may be aggregated, thus reducing datasets with thousands or even millions of observations down to one observation per cluster.

The notion of what constitutes a cluster is vaguely defined. For example, consider the points in Figure 7.1. How many clusters do you see? Your answer is probably three based on the three sequences of points. In this case, your brain has clustered the points based on a perceived visual **connectedness**.

Now consider the points in Figure 7.2. How many clusters do you see? The answer here may be more debatable. Are there two clusters, one on the left and the other on the right, or are there four clusters, two on the left and two on the right? Either way, the clusters identified are based on **proximity**.

Consider the points in Figure 7.3. You will probably agree that there are two clusters here, but what are the criteria used? Identification of the inner cluster is based on proximity. Yet if proximity is applied to points A and B, they would not be in the same cluster. The outside cluster is based on connectedness.

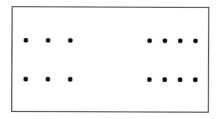

**Figure 7.2**   Proximity

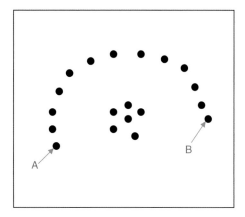

**Figure 7.3** Connectedness and Proximity

Look at Figure 7.4. How many clusters are there? Your answer is probably two. In this case the clusters are defined based on the **density** of the points. What about the other points not included in the two clusters? Are they not a part of any cluster or are they individual clusters in and of themselves? The answer to that

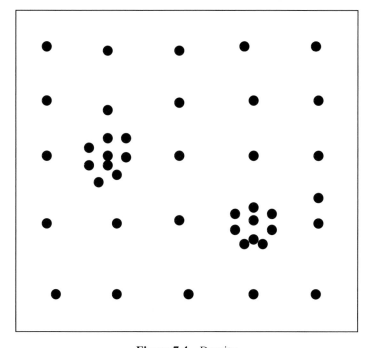

**Figure 7.4** Density

question depends more on the objective of the cluster analysis than on any predefined cluster definition.

## Algorithms for Cluster Analysis

One of the first and more simple, yet still widely used, clustering algorithms is **K-means**. Clusters are identified by the algorithm based on proximity. It uses the concept of a **centroid** which is defined as the mean of a group of points. In a dataset defined in $n$ dimensions, that is with $n$ attributes or columns, each centroid is assigned a value to each of the $n$ dimensions. Before beginning a cluster analysis using K-means, the analyst must first choose K – the number of expected clusters.

The steps of the algorithm are:

1. Randomly locate K initial centroids within the $n$-dimensional space. (Alternatively, randomly choose K observations from the dataset to serve as the initial centroids.)

2. Repeat:

   a. assign each of the observations in the dataset to the nearest centroid

   b. recompute each centroid's location as the mean of all observations assigned to that centroid until observation assignments to centroids do not change.

## Issues with K-Means Clustering Process

Although K-means is simple to understand and implement, it does have shortcomings:

- K, the number of clusters, must be set before initiating the process.

- K-means generates a complete partitioning of the observations. There is no option to exclude observations from the clustering.

- When initial centroids are randomly located, the resulting clusterings may vary from execution to execution. The end result is not deterministic.

- K-means does not handle well datasets containing clusters of varying size. In general, it will tend to split the larger clusters and may merge smaller clusters.

# Hierarchical Clustering

Another relatively simple method to perform cluster analysis is hierarchical clustering which generates a taxonomy or hierarchy of clusters. It has two alternative approaches: bottom-up and top-down.

In bottom-up hierarchical clustering, each observation is assigned to its own cluster. Repeatedly, the closest two clusters are merged until only one cluster remains or until each cluster reaches a predetermined minimum measure of cluster cohesiveness.

In top-down hierarchical clustering, we begin with one cluster containing all observations. Repeatedly, clusters are divided until all clusters reach a predefined maximum measure of cluster cohesiveness. Top-down hierarchical clustering has the additional complexity in that there needs to be a way to select the next cluster to be split, and once selected, there needs to be a way to allocate observations to the two newly created clusters. The process of top-down clustering is similar to the process of tree building presented in Chapter 4. In decision trees, the degree of homogeneity of a node is based on the single classification variable and the split of a node is based on criteria that result in the most homogeneous nodes. In cluster analysis there is no classification variable. Hence, all attributes (dimensions) must be used to compute a measure of cluster dispersion in selection of nodes to be split. This will be discussed later when we present measures of individual cluster and overall clustering quality.

Over the years numerous other methodologies have been proposed for cluster analysis. Some have enhanced the existing K-means and hierarchical proximity based methodologies, while others have focused on density or connectedness based methodologies. Self-organizing maps, which will be introduced later in this chapter, can be thought of as an enhanced K-means algorithm. For more information on these algorithms, the reader is directed to books dedicated primarily to cluster analysis.

# Measures of Cluster and Clustering Quality

Given that in cluster analysis we never know if we have "the correct answer", measures are needed to evaluate a clustering's quality. In general terms, a clustering based on proximity is valid if we have clusters that individually are **cohesive** (tightly packed around a centroid) and distinctly **separated** from the other clusters in the clustering.

A measure of cluster cohesiveness is the sum of the squared error (SSE). SSE is defined as the sum of the squared distances of each observation from the cluster's centroid. More formally it is:

$$SSE = \sum_{x \in C} dist(c, x)^2$$

where $x$ is an observation in cluster $C$ and $c$ is the cluster centroid. If all observations are tightly packed around the centroid, the SSE is relatively low. When observations are spread, the SSE is greater.

Since in a clustering, individual clusters will vary in size (number of observations), SSE will generally be larger in clusters containing more observations. To directly compare cohesiveness between clusters we compute the mean squared error (MSE) of a cluster as:

$$MSE = \frac{SSE}{m}$$

where $m$ is the number of observations belonging to the cluster. A special case to be aware of is the single observation cluster where SSE and MSE will always be zero.

To compare clusterings – that is clusterings generated by different proximity based clustering algorithms or multiple executions of the same algorithm – with respect to overall cluster cohesion we compute the total sum of the squared error (TSSE) as:

$$TSSE = \sum_{c \in CL} SSE_c$$

where $c$ is a cluster within the full clustering $CL$. Be forewarned in comparing clusterings with significantly different cluster counts, the greater the number of clusters in a clustering, the lower the TSSE. At the extreme, a clustering of one observation per cluster has a TSSE of zero. Certainly one would not expect this to be a useful clustering.

An overall measure of cluster separation is the total "between group" sum of squares (TSSB). TSSB is the sum of the squared distance of cluster centroids from the dataset overall mean (the dataset centroid) weighted by the number of observations in the cluster. It is computed as

$$TSSB = \sum_{i=1}^{K} m_i \cdot dist(c_i, c)^2$$

where $m_i$ is the number of observations in cluster $i$, $K$ is the total number of clusters, $c_i$ is the centroid of cluster $i$, and $c$ is the overall dataset centroid. When comparing clusterings, the greater the TSSB of the clustering, the better separation.

## Silhouette Coefficient

A measure that combines both cohesion and separation is the silhouette coefficient. For a single observation it is computed as

$$s_i = \frac{(b_i - a_i)}{\max(a_i, b_i)}$$

where $a_i$ is the average distance between the $i^{th}$ observation and all other observations in the same cluster; and $b_i$ is the minimum average distance of the $i^{th}$ observation to all other clusters.

The silhouette coefficient ranges in value from $-1$ to $1$. When an observation is closer on average to observations in another cluster than to observations in its own cluster, then $b_i$ is less than $a_i$ and the coefficient is negative – an undesirable result. The ideal silhouette coefficient is 1, which occurs when $a_i$ is 0 (all observations in the cluster congregate at the centroid).

To find the silhouette coefficient for a cluster, compute the average silhouette coefficient of all observations in the cluster. To find a clusterings overall silhouette coefficient, compute the average silhouette coefficient of all observations.

## Correlation Coefficient

Another measure of clustering validity that combines both separation and cohesion is the coefficient of correlation between distance and correctness. Suppose that an $m$ by $m$ distance matrix $D$ is constructed to hold distances between all observation pairings where $d_{ij}$ is the distance between observation $i$ and observation $j$. A second $m$ by $m$ indicator matrix ($C$) is constructed to hold a 0/1 indicator value reflecting whether the corresponding observations are in the same (0) or different clusters (1). See Figure 7.5. Therefore, $c_{ij}$ is 0 if observations $i$ and $j$ are in the same cluster and 1 otherwise. The coefficient of correlation for the clustering is computed as the Pearson correlation between the two matrices, pairing only values below the main diagonal (due to symmetry). In a valid clustering larger distance values will more likely be paired with 1's in the indicator matrix and lower distance values will more likely be in the same cluster (0 indicator values).

## Self-Organizing Maps (SOM)

The self-organizing map algorithm was developed by Tuevo Kohonen. It is similar in approach to K-means and its variants. The primary differences are

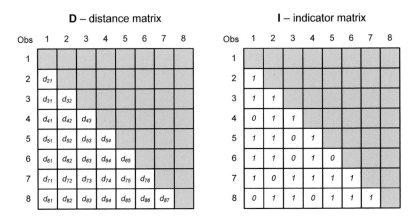

Clusters: {1, 4, 8} {2, 7} {3, 5, 6}

**Figure 7.5**    Matrices used to compute clustering correlation coefficient

that the SOM algorithm utilizes a topologically fixed centroid structure and, based on that topology, when one centroid is updated, neighboring centroids are also updated. Suggested topologies for constructing a SOM are a single-dimension linear ordering of centroids (Figure 7.6a), a two-dimensional grid (Figure 7.6b), and a two-dimensional honeycomb pattern (Figure 7.6c).

The steps of the algorithm are:

1. Select topology including number of centroids (cells).

2. Randomly initialize all centroids

3. Repeat:

   a. select next observation

   b. locate centroid closest to the observation (winning centroid)

   c. update winning centroid and other centroids in winner's neighborhood, nudging each centroid closer to the selected observation until a

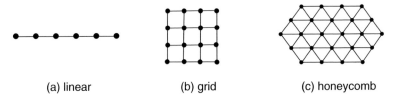

(a) linear                        (b) grid                        (c) honeycomb

**Figure 7.6**    SOM Topologies

previously determined number of training iterations have been completed or the centroid updates fail to reach a minimal change threshold.

4. Assign each observation to its winning centroid to form the clusters.

Observations are sequentially selected in step 3a above. Once the last observation has been applied, the selection process returns to the first. One cycle through the list of observations is known as an **iteration**.

For a given centroid, the magnitude of the update in step 3c above depends on its proximity to the winning centroid. The winning centroid itself is nudged the most, while those centroids immediately surrounding the winner are nudged less, and so forth, based on a Mexican hat neighborhood function. The result is that as the training process progresses, similar observations migrate toward the same or neighboring centroids. In the end, the most similar observations are assigned to the same cluster (centroid); neighboring clusters are more similar; while distant clusters are the least similar.

## Self-Organizing Maps in VisMiner

For clustering, VisMiner implements self-organizing maps. We'll try it out first with a familiar dataset iris.csv.

⏺ Open the iris.csv dataset.

Iris.csv was previously used in classification, which requires known categories in order to generate prediction models. For cluster analysis, we'll pretend that we don't already know the iris varieties and use VisMiner to see how well the SOM isolates each variety.

⏺ Since we don't want the SOM clusterer to see the variety, create a derived dataset named IrisSize that contains just the four flower measurements, but not the Variety.

⏺ Drag the SOM clusterer over the IrisSize dataset and release.

The VisMiner SOM implementation uses a grid topology extended to three dimensions. In the SOM, neighboring clusters represent somewhat similar observations, the third dimension allows nearby clusters to extend in one more direction. Hence, we are able to see observations that are similar but different in one way, appear along one dimension while observations that are similar but different in a different way appear along another dimension. The VisMiner default size is a $3 \times 3 \times 3$ grid – 27 cells.

⌲  Leave the grid size unchanged.

⌲  Check the boxes to compute all three statistics.

⌲  Click OK.

The small iris dataset processes almost immediately. Larger datasets will take more time. If processing time is an issue, you may not want to compute all statistics. Computing the silhouette coefficient can be especially time consuming.

⌲  Drag the newly created SOM:IrisSize model up to a display and drop.

⌲  Select "SOM Viewer".

The primary SOM viewer pane (Figure 7.7) represents the grid. Each cluster in the grid is drawn as a sphere whose size corresponds to the number of observations belonging to the cluster. To get an accurate assessment of the grid and the cluster locations within the grid, it helps to rotate by dragging either up and down or left and right. Note: The grid that you see when working with the datasets in this chapter may mirror or reflect images of the published figures. As the cluster analysis progresses, dominant clusters migrate to the corners of the grid. The actual ending corner for a given cluster depends on the initial random values assigned to the centroids.

The options panel (Figure 7.8) presents analysis statistics and supports some grid formatting options. In this particular clustering, the overall mean squared error (MSE) is 0.0423. This value only becomes meaningful when comparing clusterings or when comparing the MSE of an individual against the overall MSE. The clustering's coefficient of correlation is 0.516, a respectable value.

⌲  In the Transparency Options box, select the "MSE Base" radio button.

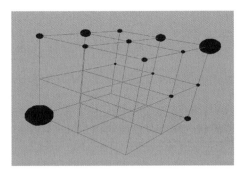

**Figure 7.7**   Iris SOM Grid

**Figure 7.8**   SOM Options Panel

The grid uses sphere transparency to indicate the validity of a cluster relative to other clusters in the clustering. When "MSE Base" is selected, the clusters with the lower (better) MSE measures are displayed as more opaque. The transparent spheres have the more undesirable higher MSEs. Notice in general that the smaller clusters are more opaque, an artifact of their small size, yet due to their size, not very useful.

Select the "Silhouette Base" radio button.

With respect to cluster quality, there is no adequate single measure. Each has its strengths. It is a good idea to consider multiple measures. MSE is a good measure of cluster cohesiveness, yet it does not consider separation. The silhouette coefficient includes both cohesiveness and separation, yet the question of how to balance between the two is quite arbitrary.

Compare the two largest clusters at the opposite corners of the grid. The cluster that is most separated from the others is more opaque, indicating a better silhouette coefficient. The same cluster was also more opaque when encoded using the MSE base. Clearly this cluster is more valid than the smaller and more transparent cluster.

Click on the larger more separated cluster located at one of the corners.

As a cluster is selected, it is highlighted using color, and the statistics (if computed) are shown in the options panel. In the case of the selected cluster, there are 50 observations in the cluster; its MSE is 0.0471, slightly worse than the overall MSE of 0.0423, and its silhouette coefficient is a very good 0.633.

☞   Ctrl-click on the second largest cluster in the opposite corner.

With two clusters selected, we can compare their qualities. The MSE of the second is higher (0.0668 versus 0.0471) and its silhouette coefficient is lower (0.054 versus 0.633). This should be no surprise since we already discovered this when comparing opacities.

Note also in the "Selected Cell Statistics" box, that each cluster is given a unique name based on its position along the X, Y, and Z directions of the grid. We will use these names later on in the tutorial.

Aside from the cluster validity and size measures, the grid provides little information to assist in understanding the clustering. To do this we need to be able to compare observations in the clusters.

☞   In the Control Center, drag the Iris.csv dataset up to a display, opening in a pane to the side of the SOM grid.

☞   Select "Parallel plot", then "Sync with SOM:IrisSize".

We are choosing to view the parent dataset Iris.csv rather than the IrisSize dataset which was used to perform the cluster analysis, because it also contains the eliminated Variety column. One objective in clustering the iris dataset was to assess the quality of the SOM clustering algorithm. We already know that there are three sub-populations in the dataset. Out of curiosity, you might wonder how the SOM clustering compares to the already known variety sub-populations.

When synchronized with the SOM clustering, the parallel plot view of the iris.csv dataset is automatically updated as clusters are selected on the grid. If you have followed the instructions carefully, you should see the two selected clusters from the outside corners of the grid presented in the parallel plot (Figure 7.9).

In the parallel plot, we see that the large isolated cluster with 50 observations is all Setosa. The SOM successfully isolated this variety. Remember, in the original dataset there were 50 Setosa, 50 Versicolor, and 50 Virginica observations. In the plot, we also see that the most distinguishing feature of this cluster is the small flower size (low PetalLength and PetalWidth values).

As a reminder that the largest cluster contains all Setosa flowers, let's give it a more meaningful name.

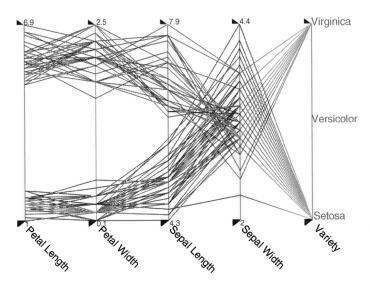

**Figure 7.9**   SOM Synchronized Plot

👉 Right-click on the largest cluster.

👉 Select "Rename cell".

👉 Give this cluster the name "Setosa".

👉 Click "OK".

The other cluster in the parallel plot contains 33 of the 50 Virginica observations. The other 17 observations must be in other, yet to be explored, clusters.

👉 In the SOM grid, click on the next largest cluster adjacent to the corner cluster with the 33 Virginica observations.

The selected cluster contains 15 observations. All but one are Virginica. We have now found 14 more of the 50 total Virginica observations. Note that the one Versicolor observation in the cluster is very similar to the other cluster members. The failure of the SOM to correctly isolate all observations is not an indictment of the algorithm, but simply due to the fact that there are overlaps in similarity between the Virginica and Versicolor varieties.

👉 Ctrl-click again on the corner cluster containing the 33 Virginica observations.

The two adjacent clusters should now be represented in the parallel plot. How do they compare?

⌘  In the parallel plot, check the "Show Means" box.

The smaller of the two clusters represents overall smaller flowers. Because the mean plot lines of the two clusters are roughly parallel, the ratios between measures within each of the clusters appear to be similar.

As a matter of practice, when comparing clusters in a synchronized parallel plot, it is suggested that the "Show Means" box be checked.

## Choosing the Grid Dimensions

The choice of grid dimensions when building a SOM clustering depends on the objective of that clustering. Smaller grids with fewer total cells force what would normally be adjacent cells into the same cell. Hence, if the objective is to isolate a few large clusters that can be individually studied using additional data mining methodologies, choose a small grid size.

Consider the cluster analysis just completed for the iris dataset with the $3 \times 3 \times 3$ grid. There were 15 total clusters generated. Six clusters contained three or fewer observations.

⌘  Redo the Iris clustering, choosing a grid size of $2 \times 2 \times 1$.

⌘  Open Iris.csv in a parallel plot synchronized with the SOM. Use the parallel plot to evaluate each of the three clusters generated.

⌘  Compare statistics between the two clusterings. (See Table 7.1, rows one and three.)

⌘  Again redo the Iris clustering choosing a grid size of $2 \times 2 \times 2$. Evaluate and compare.

**Table 7.1**   Grid Statistics

| Grid Dimensions | No. Clusters | No. Small Clusters (<4 obs) | MSE | Correlation |
|---|---|---|---|---|
| $2 \times 2 \times 1$ | 3 | 0 | 0.1060 | 0.704 |
| $2 \times 2 \times 2$ | 5 | 1 | 0.0872 | 0.648 |
| $3 \times 3 \times 3$ | 15 | 6 | 0.0423 | 0.516 |

Comparing the clusterings using MSE, a measure of cohesion only, the $3 \times 3 \times 3$ clustering would be considered best. Keep in mind, however, that smaller clusters tend to have lower MSEs. Comparing the clusterings using the coefficient of correlation, a measure of both cohesion and separation, the $2 \times 2 \times 1$ clustering comes out on top.

If the objective of the clustering is to gain a better understanding of the data, all clusterings provide some insight. In the $3 \times 3 \times 3$ clustering, the separation of the Setosa cluster indicates that these flowers are quite distinct from the others. The numerous adjacent clusters near the opposite corner reflect a difficulty in categorizing both the Versicolor and Virginica flowers based on petal and sepal size measurements. The larger merged cells in the smaller clusterings again reflect the difficulty in distinguishing between Versicolor and Virginica.

## Advantages of a 3-D Grid

To illustrate the advantages of a 3-D grid, let's evaluate the voting records of US congressional representatives. In 2004, the US House voted on 50 proposed acts. The dataset SelectedVotes.csv contains the voting records for representatives on the acts that passed with less than a 90% majority. In other words, those votes in which almost all representatives were in agreement have been removed. There is one observation per representative. For each act, the name of the column is the name of the act. A "yes" vote is recorded as 1 and a "no" vote as 0. If the representative did not vote on the act in question, it is assigned a value of 0.5. In addition to the voting record, the dataset also contains RepName, State, and Party.

↪ Open the SelectedVotes.csv dataset.

↪ Create a derived set named "votes" with the vote columns only. That is, exclude the Party, RepName, and State columns.

↪ Create a $3 \times 3 \times 3$ clustering of "votes".

↪ Open the resulting model in the SOM viewer.

↪ Open a tabular presentation of SelectedVotes.csv that is synchronized with the "votes" SOM.

The SOM cube (Figure 7.10) contains two large clusters on opposite corners. You can probably guess which groups these clusters represent. They are those that consistently voted the party line. The largest cluster contains only Republicans, while the other contains only Democrats.

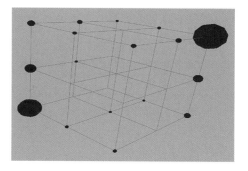

**Figure 7.10**   SOM Vote Clustering

To verify, click on each of the two clusters, then review in the synchronized data table. (You may want to stretch the "RepName" column header to make room for the full name.)

The clusters adjacent to the two dominant clusters contain representatives that usually voted the party line yet diverged at times. This is where the advantage of the 3-D SOM becomes evident. There are three clusters adjacent to the "Republican" cluster. Each identifies a different pattern in the way they strayed.

The clusters in the cells distant from the two dominant clusters represent the more independent voters. They did not vote the party lines, yet given the distances between some of these independent clusters, their patterns of deviation varied. Out of curiosity, you may want to click on each of these clusters to see in the synchronized data table who these representatives are and the states they come from.

☞ To compare the voting records between clusters, close the data table and open in its place a synchronized parallel plot of the "votes" dataset.

☞ Use the "Ctrl-click" option of the SOM to select multiple clusters to compare.

☞ The default plot is ugly and difficult to interpret. Check "Show Means" to get a more meaningful plot.

Without the three-dimensional grid, we would not have had that extra dimension along which clusters could vary from the dominant corner clusters.

## Extracting Subsets from a Clustering

The dataset OutdoorCustomerSales.csv contains a sampling of sales data for online sales of Big Box Merchandising, a large warehouse retailer. The Purchases column contains the total online purchases for the year of outdoor sports equipment by each of the sampled Big Box customers. There are 5,000 customers in the sample.

Big Box would like to direct some targeted magazine based promotions at customers most likely to respond. To assist in their analysis, Big Box surveyed each of these 5,000 customers to determine which magazines they read. Their responses are in the dataset MagazineReadership.csv. For each magazine surveyed (Cosmopolitan, Jet, Newsweek, PCMag, People, and Sports Illustrated), readership is coded as 1 and non-readership as 0. There is no in-between "occasionally" option.

⇨ Join the two datasets by dragging MagazineReadership.csv and dropping it on OutdoorCustomerSales.csv; then select "Join Datasets".

⇨ In the Join Parameters dialog, check CustomerID as the matching column in each of the datasets.

⇨ Click "OK".

⇨ Create a derived dataset named "data" from the resulting joined dataset. Include the following columns in the dataset: Cosmopolitan, Jet, Newsweek, PCMag, People, SportsIllustrated, and Purchases.

⇨ Create a derived dataset named "readership" from "data" to be used for the cluster analysis. It should include all the six magazine columns, but not Purchases.

⇨ Create a $3 \times 3 \times 3$ clustering of readership.

⇨ View the SOM:readership model in the SOM viewer.

⇨ Open "data" in a parallel plot synchronized with the SOM:readership clustering.

⇨ Drag the Purchases column all the way to the right to keep it separated from the six magazine readership columns.

⇨ Check "Show Means".

⇨ Click on the largest cluster.

The largest cluster contains 2,434 observations. In the parallel plot, notice that readership is zero for all six magazines. This is the cluster representing customers that read none of the six surveyed magazines.

☞   Right-click on the large cluster; select "Rename cell". Give it a more meaningful name: "non-readers".

☞   To see the names, check "Show Names".

☞   One by one, click on each of the other clusters. Give them a name, based on the magazine readership that you see in the synchronized parallel plot. For example, the cluster in the opposite corner shows high readership for Newsweek and People. Name it "NWPeople". Use the following names for the other columns:

People – the cluster with 820 observations.

PCMag – with 132 observations.

Jet – the smallest cluster with 76 observations.

Cosmopolitan – with 123 observations.

Newsweek – with 324 observations.

PeopleSI – with 218 observations.

SI – with 464 observations.

☞   Use the "Ctrl-click" option on each cluster to simultaneously select all nine clusters.

☞   Right-click on any cluster; select "Make dataset from all selected".

☞   When prompted to select columns, leave all checked. Press "OK".

☞   In the Control Center, view the summary statistics for the newly created "selected" dataset.

The "selected" dataset contains all of the original dataset columns plus a new column – ClusterName. This column is automatically created by the "Make dataset from all selected" option in the SOM viewer when multiple cells are extracted. The cardinality of this column is 9 – one for each selected cell. The number of rows in the dataset is still 5,000 – the original sample size.

To summarize, at this point we have clustered the customers according to their magazine readership, then added a column, ClusterName, to each customer row with the name of the cluster to which it was assigned.

☞   Right-click on "selected"; then select "Aggregate rows".

↩ In the dialog that opens, aggregate on ClusterName; check Purchases for aggregation; check "Avg".

↩ Name the aggregation "AvgPurchases"; click "OK".

↩ View the newly created AvgPurchases.csv dataset in a parallel plot.

Which cluster of customers has the highest average purchase amount? Which cluster has the lowest? Based on the results, which magazine (or magazines) should receive the heaviest advertising during the upcoming promotion?

Notice that via aggregation of clusters we have achieved the last objective of cluster analysis, reducing the 5,000 observations down to nine.

## Summary

Cluster analysis is the process of grouping similar observations. It has three primary purposes:

- to identify and isolate sub-populations
- to improve understanding of the dataset relationships
- to aggregate observations for reduction purposes.

The result of a cluster analysis is called a clustering. The quality of a clustering is defined with respect to its cohesiveness within clusters and separation between clusters.

VisMiner implements a three-dimensional version of the self-organizing map (SOM) algorithm for cluster analysis. It provides for synchronized visualizations of the overall clustering, individual clusters, and clusters selected for comparison. It also supports extraction of clusters for use by other data mining algorithms.

# Appendix A

# VisMiner Reference by Task

## Dataset Preparation

Handle missing values

VisMiner does not support mining of datasets with any missing values. They are represented in the Control Center by a dataset icon overlaid with a red triangle. Before proceeding with any data exploration or mining tasks all missing values must be handled. To do this:

1. Right click on dataset containing missing values.

2. Select "Handle missing values".

3. Specify handling option for all attributes with missing values.

Outlier detection and isolation

The search for outliers in a dataset takes time, but is facilitated by the data exploration viewers of VisMiner. For a thorough search, complete all of the following:

1. View Summary Statistics for dataset. (Right-click on dataset in Control Center.)

   a. Look at minimum and maximum values of each numeric attribute; check that they fall within acceptable range. For example, an attribute GradePointAverage should not have a minimum value less than zero or a maximum value greater than four.

*Visual Data Mining: The VisMiner Approach*, First Edition. Russell K. Anderson.
© 2013 John Wiley & Sons, Ltd. Published 2013 by John Wiley & Sons, Ltd.

b. For nominal attributes, hover over the cardinality value. When hovered, if there are not too many different values, the unique values will be displayed in a small pop-up box. Check that all values are acceptable.

2. Use either the histogram or parallel coordinate plot to view the distributions of each attribute. Look for outliers at the extremes of numeric attributes.

3. Some outlying observations may not contain extreme values with respect to any single attribute, yet may still be outliers because they do not fit a relationship pattern between two somewhat correlated attributes. To detect, open the dataset in both a correlation matrix and scatter plot. The scatter plot will automatically synchronize with the correlation matrix. In the correlation matrix, click on each of the correlated attribute pairings to view in scatter plot. Look for observations that do not fit the pattern of correlation. For example, suppose that a dataset contains sales quantities of winter coats by city. Included in the dataset are location attributes of the city (latitude and longitude). The normal pattern would be for sales of winter coats to increase as the latitude increases. A southern city with high winter coat sales would be an outlier.

4. Use the parallel coordinate plot to restrict observations to only those having a selected category value. It is possible that outliers will be visible with respect to a single category, that are not visible when all observations are viewed. For example, an observation with an attribute value of Pregnant = Yes would not be visibly detected when viewing all observations, yet would stand out when viewing only Gender = Male observations.

5. When datasets contain attributes that can be derived from other attributes, at least one of the attributes is redundant. Add a computed column to the dataset that uses other attributes to duplicate the derived attribute. Use the parallel coordinate plot or scatter plot to compare the newly computed column to the derived attribute. If they don't match, there is a problem in either the derived attribute or one of the attributes used to compute its value.

## Outlier isolation

Outliers, once detected using any of the VisMiner viewers, are best isolated or eliminated using the parallel coordinate plot:

1. Eliminate observation(s) based on a single attribute outlier by dragging the filter slider of the attribute past the outlying observation(s). Right-click on slider to "Make dataset from filter".

2. To remove outlying observations based on patterns between two or more attributes:

   a. Isolate observation(s) using the filter sliders.

   b. Right-click on slider to "Make dataset from filter".

   c. In the Control Center,
      i. drag and drop the newly created dataset of the isolated outlying observation(s) onto the parent dataset

      ii. select "Create dataset from difference" to generate a dataset containing all but the isolated observation(s).

## Dimension reduction

Dimensions (attributes) should be removed from a dataset when they are not considered applicable or of value with respect to a planned data mining task. Some attributes to be eliminated are obvious candidates. For example, customer account numbers would not be expected to contribute to any planned pattern analyses. Use features of the Control Center to eliminate these attributes. Other attributes should be eliminated if, after review, they appear unrelated or weakly related to the planned data mining task. The correlation matrix and parallel coordinate plot viewers can help to first identify then eliminate these attributes.

   Attribute elimination using the Control Center:

1. Selectively removing columns:

   a. Right-click on the dataset.

   b. Select "Create derived dataset".

   c. Check only those columns to be included in new dataset.

2. Creating a single summary attribute to replace multiple attributes:

   a. While in the "Create derived dataset" option, create a "computed column" based on the attributes to be summarized. For example, create a column TotalSales as the sum of SalesX, SalesY, and SalesZ.

   b. Remove (leave unchecked) the newly summarized attributes from the derived dataset.

   Attribute elimination using the correlation matrix:

1. Ctrl-click on unwanted attribute names.
2. Click "Create Subset" button to save.

Attribute elimination using the parallel coordinate plot:

1. Ctrl-click over axes of unwanted attributes.

2. Right-click on filter slider to "Make dataset from filter".

Observation reduction

1. Sampling – To create a random sample of observations using the Control Center:

   a. right-click on dataset

   b. select "Create derived dataset"

   c. select columns to be included

   d. enter desired sample size in "Rows for new derived set".

2. Aggregation – The Control Center allows you to create datasets that are summaries (aggregations) of rows. It requires specification of the grouping column(s) and the aggregations to be computed (row count, sum, average, minimum, and maximum). Its functionality is the same as the SQL "select" statement with a "group by" clause. To aggregate:

   a. right-click on dataset

   b. select "Aggregate rows"

   c. select "Aggregate On" columns which will be used for grouping

   d. select aggregation functions.

3. Identification and extraction of sub-populations – For visual identification and extraction use either the parallel coordinate plot or the location plot. For automated processing use the SOM clusterer.

   a. Parallel coordinate plot
      i. Adjust the filter sliders to visually isolate a sub-population.

      ii. Right-click any slider and select "Make dataset from filter".

   b. Location plot
      i. Use range sliders and category filters to restrict visible observations.

      ii. Click "Subset" button to extract only visible.

   c. SOM clusterer
      i. Drag SOM clusterer over dataset and drop.

      ii. Select SOM options.

iii. Drag newly created clustering up to display and drop.

iv. Select SOM Viewer.

v. Right-click on any cluster cell to "Make dataset from cluster".

## Creating training, validation, and test sets

Datasets to be used by modelers allow the partitioning of the dataset into training and validation partitions. Observations in the training partition are used to actually build the model. Rows in the validation partition are used during model construction to avoid overtraining, and after construction to assess the model's generalizability. Once a model is constructed, other datasets having the same structure (same column names and types) may be applied to the model to generate predictions. These datasets, applied later, are herein referred to as "test" sets.

Use the Control Center to partition into training and validation sets:

1. Right-click on dataset to be partitioned.

2. Select "Create derived dataset".

3. Check the "Columns to Include".

4. The number of training rows is specified in "Rows for new derived set".

5. The number of validation rows is specified in "Rows for validation set".

6. Once created, drag and drop modeler over dataset to build model.

Apply a test dataset:

1. Drag a compatible dataset (same column names and types) over the model and drop.

2. Select "Test model performance" to create a test set that can be viewed in the confusion viewer, ROC viewer or other model viewers.

3. Select "Generate Predictions" to create a new dataset with a new "predicted values" column.

## Balancing/stratified sampling

In many datasets used for classification, the frequency of positive response observations is small relative to the frequency of the negative response observations. These imbalances can make it difficult for the modeler to successfully extract rules for positive response prediction. To get around this problem, it is advisable to create a new dataset containing more balanced positive and negative frequencies. The following steps accomplish this:

1. Use the parallel plot to create a subset of positive response observations.
2. Use the parallel plot to create a subset of negative response observations.
3. In the Control Center create a sampled subset of the negative response dataset containing the same number of rows as the positive response dataset.
4. Drag the sampled negative response set over the positive response set and drop.
5. Select "Merge datasets".
6. Right-click on the newly merged dataset; select "View/Edit names and notes" to give it a more meaningful name.

Joining datasets

Frequently the data needed for an analysis is found in multiple datasets. For example, a business may generate its own sales data that, for a complete analysis, needs to be combined with population data from the census bureau. A row by row combination of such data is known as an **equi-join**. It requires common identifying attributes in each dataset used to match rows in one dataset to rows in the other. To do this:

1. open both datasets
2. drag one dataset over the other and drop
3. select "Join datasets"
4. in the join dialog, identify the columns in each dataset that are to be used to match rows.

## Data Exploration

Dataset overview

In the Control Center, right-click on a dataset to view its summary statistics. The summary includes: number of observations, attributes names (columns), and corresponding data types along with summary statistics for each attribute. For datasets with missing values, information on quantity and location of the missing values is also summarized.

Distribution assessment

Control Center summary statistics – contains minimum, maximum, mean, and standard deviations for all numeric attributes. For discrete attributes (text and integer), it provides cardinalities (number of unique values) with distribution counts shown in pop-up as the cardinality cell is hovered.

- Histogram – selectively creates histograms of each attribute in dataset.

- Parallel coordinate plot – observation densities are both color encoded and plotted as heights along Z dimension.

- Scatter plot – distributions become visible when same attribute is chosen for both X and Y axes.

Pattern/relationship search

The ultimate purpose of data mining is to find previously unknown relationships or patterns between attributes. The search begins during initial exploration when patterns of potential interest are first identified. In later data mining steps, a more in-depth analysis of these patterns is conducted using the data mining algorithms. Viewers supporting pattern search during initial exploration include:

- Correlation matrix – displays color encoded correlations between all attribute pairs. It visually draws attention to related attributes.

- Correlation matrix synchronized with scatter plot – when the same dataset is viewed using both the correlation matrix and scatter plot, the axes of the scatter plot are synchronized with attribute pair selections in the correlation matrix. Use the correlation matrix to quickly identify correlated attributes, then methodically click on each cell of interest in the correlation matrix to show the relationship in the scatter plot. With each selection you may want to add a third Z axis attribute to the plot for more complex pattern searches. If the objective is classification, select the classification attribute as the scatter plot "Category" attribute for color encoding.

- Parallel coordinate plot – although the PCP is best used for subset recognition and extraction, it also supports pattern searches. Crossing line patterns between adjacent axes represent inverse correlation, while nearly parallel line patterns represent direct correlations. To assess patterns between non-adjacent axes, drag one axis toward the other until they are adjacent. If the objective is classification, create filters for each of the classification attribute values. Note which attributes best discriminate between filters. If there are too many observations, and the plot becomes so cluttered that it inhibits accurate interpretation, check the "Show means" box.

- Boundary plot – is most useful for detecting patterns with respect to political boundaries. Current boundaries supported include US state, US county, three-digit zip code, and five-digit zip code. If your data is summarized, or may be summarized via aggregation, by any of these political boundaries, then use the boundary plot to visualize patterns based on geographic location.

- Location plot – is most useful for detecting patterns with respect to geographic point locations encoded via latitude and longitude. If your dataset contains location information, such as an address, but does not include latitude and longitude, you can add latitude and longitude using external geocoding tools or join your dataset with datasets containing these values.

## Model Building – Algorithm Application

To create a model using one of the available data mining algorithms, drag the modeler (data mining algorithm) over the target dataset and drop. Before doing this, however, be sure that the dataset is ready for processing. The modeler will use all observations and all attributes contained in the dataset. If you don't want to use all of the data, first create a subset of the data, eliminating any unnecessary or unwanted attributes and observations.

Choose a modeler based on the objectives of your data mining and the capabilities of the modelers. The features of the available modelers are summarized in Table A.1. They are divided into three categories: cluster analysis, classification (prediction of nominal value), and regression (prediction of numeric value). Cluster analysis is oriented more toward dataset preparation (sub-population extraction) than a data mining end point. When conducting classification or regression modeling, it is a good idea to apply multiple modelers to compare the performance results of each. No single modeler works best across all datasets.

## Model Evaluation

Once generated, data mining models should be studied and evaluated from two perspectives:

- How well does the model performs with respect to training, validation, and test datasets?

- What is the nature of the relationships between inputs and the output variable?

The evaluation approach employed varies with respect to the data mining objective (classification, regression, or cluster analysis) and the algorithm used to build the model.

**Table A.1**  VisMiner Modelers

| Modeler/Purpose | Advantages | Limitations and Weaknesses |
|---|---|---|
| SOM clusterer | • Automatically creates subsets of data observations<br>• Adjacent clusters are similar clusters<br>• Can be multidimensional | • Clustering based on Euclidian distance only<br>• Does not provide hierarchical clustering<br>• Clusters generated are non-uniform in size<br>• Nominal data limited to cardinality $\leq 30$<br>• Clustering may vary, depending on input row sequence |
| ANN classifier | • Can be trained to fit almost any data whether linear or curvilinear<br>• Readily detects interaction effects between inputs | • Will overfit if not monitored closely<br>• Structure of model difficult to interpret<br>• No available tests for significance available<br>• May settle in sub-optimal locations<br>• Given random starting location, results may vary from execution to execution<br>• Nominal data limited to cardinality $\leq 10$ |
| Decision tree (classification) | • Results are easily visualized and interpreted<br>• Fast execution | • Performance of the model may not be as good as other classifiers<br>• Nominal data limited to cardinality $\leq 30$ |
| Support vector machine (classification) | • Can be trained to fit almost any data | • Frequently overfits<br>• CPU intensive; with the same dataset, will take longer than other modelers<br>• Structure of model difficult to interpret<br>• Nominal data limited to cardinality $\leq 10$ |

*(continued)*

**Table A.1**   (*Continued*)

| Modeler/Purpose | Advantages | Limitations and Weaknesses |
|---|---|---|
| Linear regression | • Simple and quick algorithm<br>• Model easy to interpret<br>• Well defined measures of significance | • Linear only<br>• No interaction between inputs<br>• Nominal data having cardinality $\leq 10$ |
| Polynomial regression | • Support for linear and limited curvilinear<br>• Simple and quick algorithm<br>• Model easy to interpret<br>• Well defined measures of significance | • Additive model – no interaction between inputs<br>• Nominal data limited to cardinality $\leq 10$ |
| ANN regression | • Can be trained to fit almost any data<br>• Readily detects interaction effects between inputs | • Will overfit if not monitored closely<br>• Structure of model difficult to interpret<br>• No available tests for significance available<br>• May settle in sub-optimal locations<br>• Given random starting location, results may vary from execution to execution<br>• Nominal data limited to cardinality $\leq 10$ |

Model performance

Most measures of model performance may be computed using any of the three applicable datasets – training, validation, and test. These sets can also be used to compare actual outputs to predicted outputs.

For the test set only, the performance measures are not automatically computed. The dataset must first be applied to the model. (Drag and drop test dataset on model, then choose "Test model performance".)

- Classification

  1. Classification error rate
     a. Compare error rate of model to baseline error rate which is one minus the rate of the most frequently occurring class. For example, if the most frequently occurring class is found in 52% of the observations, then a model prediction error rate of 40% would be an improvement over the baseline error rate of 48%. However, if the rate of the most frequently occurring class is 95%, then a model error rate of 10% would be worse than the baseline error rate of 5%.

  2. View model error rates using the confusion viewer, the ROC viewer, and the class model viewer.

  3. False positive and false negative error rates – available in the confusion viewer. Depending on the intended model application, the costs of the different types of errors may be quite different. If one error type is more costly than another, focus on that type of error.

  4. Area under curve (AUC) – available in ROC curve viewer. Maximum AUC is 1.0. The closer to 1.0 the better.

  5. Model lift – available within ROC curve viewer. Represents error rate found when only the top n% of the observations are chosen.

  6. Model applications costs – available within ROC curve viewer. Allows user to apply monetary costs to compute benefits of model with respect to false positive and false negative errors.

- Regression

  1. $R^2$ – measure of regression fitness. Any value greater than zero is an improvement on the baseline model (output attribute mean). Available in the regression model viewer and the regression summary where applicable.

  2. F-statistic and P-value – statistical measures of goodness-of-fit with respect to the regression as a whole and to input coefficients. Available for linear and polynomial regressions only via regression summary.

- Self-organizing map
  Statistics computed for dataset clusterings measure cluster cohesiveness (how similar are the observations within a cluster) and separation (how distinct is a cluster from other clusters in the clustering set). All are available in the SOM viewer.

  1. Mean squared error (MSE) – a measure of cluster cohesion. The MSE magnitude is only meaningful relative to the MSE of other clusters within the clustering or other clusterings of the same dataset.

  2. Silhouette coefficient – a combined measure of both cohesion and separation. Its range is $[-1.0, 1.0]$, where $-1.0$ is the worst possible and $1.0$ is the best possible.

  3. Correlation coefficient – another combined measure of both cohesion and separation. Its range is $[-1.0, 1.0]$, where $-1.0$ is the worst possible and $1.0$ is the best possible.

## Model relationships

Study the relationships between input attributes and the output to evaluate the strength and nature of the relationship. Also look for interactions between inputs. Interactions occur when changes in value of one input attribute affect the nature of the contribution to output of a second input attribute. For example, the presence of a formal dining room may not add as much value to a small house as to a large house.

If studied carefully, the relationships will provide insights into the functioning of the world being modeled.

1. Surface models of the classification surface viewers and the regression model viewers depict strength of relationships between inputs and outputs.

2. The shape of the curves at opposite edges of a surface is an indicator of interaction between inputs if those curve shapes are different.

3. Tree graphs available for the decision tree classifier provide insights into the importance of inputs.

   a. Tree branching attributes at the top of the tree provide greater differentiation between output class values than those lower on the tree.

   b. The presence of attributes on one branch of a tree that do not exist on another are an indicator of input interactions.

4. For linear and polynomial regression models only, the coefficients of the regression summary directly represent input contributions to the output value.

# Appendix B

# VisMiner Task/Tool Matrix

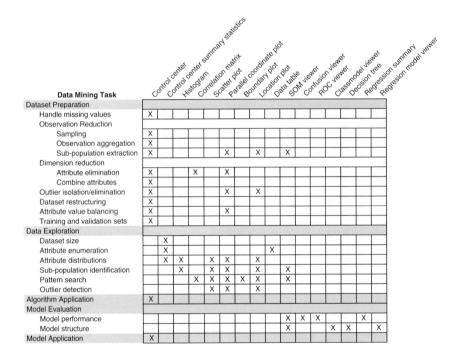

| Data Mining Task | Control center | Control center summary statistics | Histogram | Correlation matrix | Scatter plot | Parallel coordinate plot | Boundary plot | Location plot | Data table | SOM viewer | Confusion viewer | ROC viewer | Classmodel viewer | Decision tree | Regression summary | Regresion model viewer |
|---|---|---|---|---|---|---|---|---|---|---|---|---|---|---|---|---|
| **Dataset Preparation** | | | | | | | | | | | | | | | | |
| Handle missing values | X | | | | | | | | | | | | | | | |
| Observation Reduction | | | | | | | | | | | | | | | | |
|   Sampling | X | | | | | | | | | | | | | | | |
|   Observation aggregation | X | | | | | | | | | | | | | | | |
|   Sub-population extraction | X | | | | X | | X | | X | | | | | | | |
| Dimension reduction | | | | | | | | | | | | | | | | |
|   Attribute elimination | X | | X | | X | | | | | | | | | | | |
|   Combine attributes | X | | | | | | | | | | | | | | | |
| Outlier isolation/elimination | X | | | | X | | X | | | | | | | | | |
| Dataset restructuring | X | | | | | | | | | | | | | | | |
| Attribute value balancing | X | | | | X | | | | | | | | | | | |
| Training and validation sets | X | | | | | | | | | | | | | | | |
| **Data Exploration** | | | | | | | | | | | | | | | | |
|   Dataset size | | X | | | | | | | | | | | | | | |
|   Attribute enumeration | | X | | | | | | | X | | | | | | | |
|   Attribute distributions | | X | X | | X | X | | X | | | | | | | | |
|   Sub-population identification | | | X | | X | X | | X | | X | | | | | | |
|   Pattern search | | | | X | X | X | X | X | | X | | | | | | |
|   Outlier detection | | | | | X | X | | X | | | | | | | | |
| **Algorithm Application** | X | | | | | | | | | | | | | | | |
| **Model Evaluation** | | | | | | | | | | | | | | | | |
|   Model performance | | | | | | | | | | | X | X | X | | X | |
|   Model structure | | | | | | | | | | X | | | X | X | | X |
| **Model Application** | X | | | | | | | | | | | | | | | |

*Visual Data Mining: The VisMiner Approach*, First Edition. Russell K. Anderson.
© 2013 John Wiley & Sons, Ltd. Published 2013 by John Wiley & Sons, Ltd.

# Appendix C

# IP Address Look-up

## IP Address for VisSlave When Using One Computer

When using the same computer to run the Control Center and VisSlave, whether it has one or multiple displays, use the localhost IP address (127.0.0.1).

## IP Address for VisSlave When Using Multiple Computers

When using VisSlave on a computer different from the one running the Control Center, you must find the IP address of the computer running the Control Center. Use the following steps to accomplish the task:

- Using the computer which is running the Control Center, type "cmd" in the "Search programs and files" box on the Start menu.

- Press Enter.

- A DOS dialog box will appear.

- Type "ipconfig" (make sure that it is all one word).

- Press enter and the box will list the IP address/addresses of the computer (Figure C.1) along with other network related information.

*Visual Data Mining: The VisMiner Approach*, First Edition. Russell K. Anderson.
© 2013 John Wiley & Sons, Ltd. Published 2013 by John Wiley & Sons, Ltd.

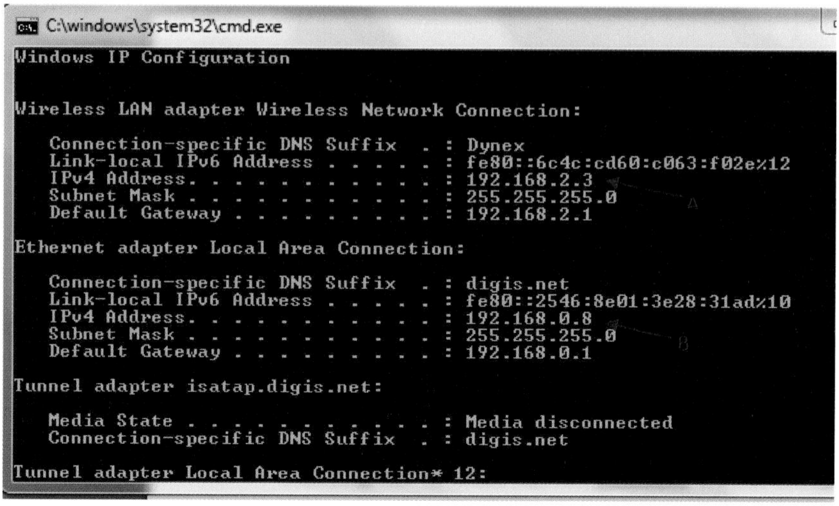

```
C:\windows\system32\cmd.exe

Windows IP Configuration

Wireless LAN adapter Wireless Network Connection:

        Connection-specific DNS Suffix  . : Dynex
        Link-local IPv6 Address . . . . . : fe80::6c4c:cd60:c063:f02e%12
        IPv4 Address. . . . . . . . . . . : 192.168.2.3
        Subnet Mask . . . . . . . . . . . : 255.255.255.0
        Default Gateway . . . . . . . . . : 192.168.2.1

Ethernet adapter Local Area Connection:

        Connection-specific DNS Suffix  . : digis.net
        Link-local IPv6 Address . . . . . : fe80::2546:8e01:3e28:31ad%10
        IPv4 Address. . . . . . . . . . . : 192.168.0.8
        Subnet Mask . . . . . . . . . . . : 255.255.255.0
        Default Gateway . . . . . . . . . : 192.168.0.1

Tunnel adapter isatap.digis.net:

        Media State . . . . . . . . . . . : Media disconnected
        Connection-specific DNS Suffix  . : digis.net

Tunnel adapter Local Area Connection* 12:
```

**Figure C.1**  IP Address

The needed IP address is found on the "IPv4 Address" line. In some instances a computer may have multiple network connections. For example, it may have both a wired (B) and a wireless (A) connection. Either address will work. Enter this number as prompted by VisSlave for the Control Center IP address.

# Index

*Visual Data Mining: The VisMiner Approach*, First Edition. Russell K. Anderson.
© 2013 John Wiley & Sons, Ltd. Published 2013 by John Wiley & Sons, Ltd.